Annals of Solitude

Annals of Solitude

—— A Year in a Hut in the Arctic ——

Stephen Pax Leonard

RESOURCE *Publications* • Eugene, Oregon

ANNALS OF SOLITUDE
A Year in a Hut in the Arctic

Resource Publications
An Imprint of Wipf and Stock Publishers
199 W. 8th Ave., Suite 3
Eugene, OR 97401

www.wipfandstock.com

PAPERBACK ISBN: 978-1-6667-3828-5
HARDCOVER ISBN: 978-1-6667-9878-4
EBOOK ISBN: 978-1-6667-9879-1

04/21/22

For Olga

Simple in means, but rich in goals.

—Arne Naess in a discussion of
"how we should live our lives"

Contents

An Explanation

For many years, I had wanted to leave our congested, complicated world behind and experience the joys of a simple life living as close to nature as possible. I lived in a hut on my own for a year with no running water or waste drainage. I had electricity and a temperamental oil heater. Via a satellite, I had access to the Internet once a week on a Wednesday morning for two hours but most weeks I forgot which day it was and missed my slot. Although I lived on my own, I did not move to the Arctic to be alone. I lived amongst a small community of hunters who like me also saw the appeal of the simple, pre-industrialized life. I chose the most remote, traditional community I could find in the north-west corner of Greenland. Savissivik: one hundred and twenty miles from the nearest village, no roads, no machines of any kind, population forty. Winter temperatures as low as minus forty-five degrees Celsius. Only way in and out: dog sled across the sea ice or the weekly helicopter (normally canceled in the winter months). In short: bliss.

Simple household chores became my daily liturgy. I collected ice for drinking and washing, wrestled with the heater, trained a small dog team and watched the seasons slip by. I discovered the spiritual appeal of the frozen world; the anonymous white bits at the top of the globe. I witnessed first-hand how our actions 'down there' in the 'civilized' world threatened to literally change the color of the map. I knew months of darkness, months of constant sunlight, mistrust, freedom, adventure, happiness, a catalogue of frustrations and finally the shock of my previous life.

I limited myself to seventy-five kilograms (one hundred and sixty-five pounds) of luggage. I decided there could be room for only one book. I took *The Penguin Book of English Verse* (all 1140 pages of it) for I could think of few other books that condensed so much meaning into one tome. Every evening, snuggled up in my sleeping bag, I read aloud some poems.

AN EXPLANATION

Every day I wrote a diary, even when sat in twenty-four-hour darkness and my achievements amounted to 'Smashed up ice. Scraped candle wax from table'. This is the record of my journal for that year.

S.P.L

Map: The Arctic and north-west Greenland

Map of the Arctic, showing north-west Greenland

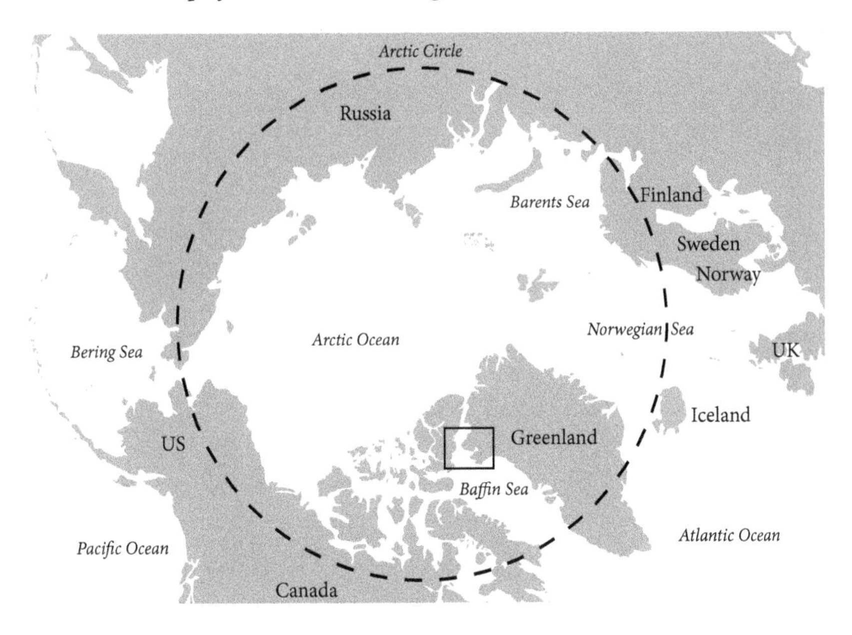

Map of north-west Greenland, showing inhabited and abandoned settlements

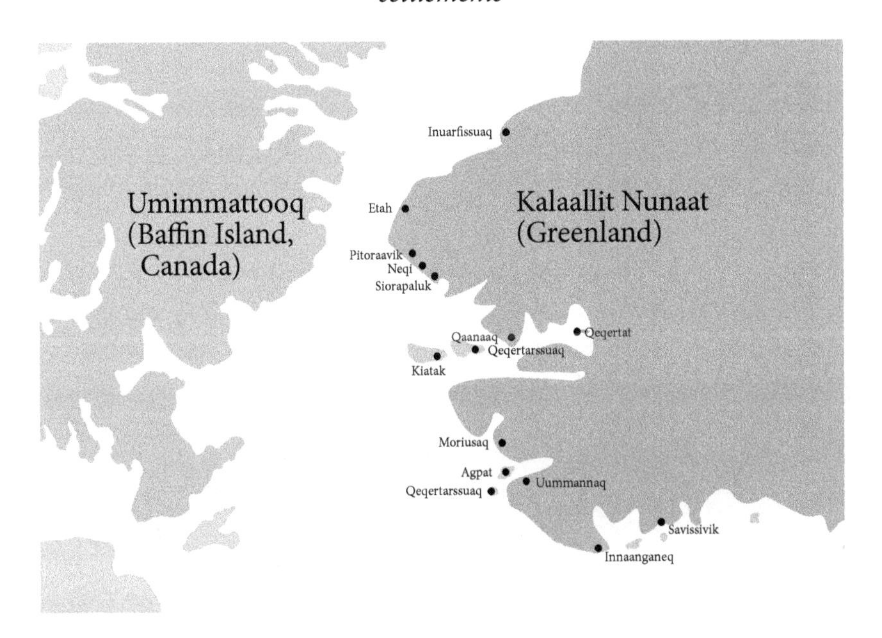

Acknowledgments

I would like to thank Matt Wimer at Wipf and Stock for believing in this project from the beginning and for his efficiency and help in bringing it to fruition. In addition, there are some other people and institutions that I would like to thank. Without their support, guidance and love, this adventure and the subsequent preparation of the manuscript would have been a hapless task. In no particular order, they are as follows:

Olga
Ibbi Qaavijaq
Ane-Sofie Imiina
David Qujaukitsoq
Moscow State Linguistic University
Fabrizia Preciosi
Peter Harris
Qaaqutsiaq Qujaukitsoq
Exeter College, Oxford
Tatiana Yatsunova
Trinity Hall, Cambridge
Rev'd Dr Stephen Plant
Professor Marklen Konurbaev
Dr Therese Feiler
Dr Alexander O'Hara

Chapter 1: *On Discovery*

Upon a great adventure he was bond,
That greatest Gloriana to him gave,
That greatest Glorious Queene of Faerie lond,
To winne him worship, and her grace to have,
Which of all earthly things he most did crave;
And ever as he rode, his hart did earne
To prove his puissance in battell brave
Upon his foe, and his new force to learne;
Upon his foe, a Dragon horrible and stearne

(from *The Faerie Queene* by Edmund Spenser)

August 20

My first sight of Greenland was the greying and thinning ice sheet. It looked tired, grown melancholic with age, mumbling chapters of disregard. Dotted all over the ice lay brilliant, sapphire blue melt-water lakes, some of them a mile or so across. I felt a sense of remorse; the lakes symbolized surely a collective human failure, a tragic testament of the all too rapidly changing natural environment.

After four and a half hours of flying up the West Greenlandic coast with nothing but bare rock, meandering glaciers and icebergs littering the fjords, we came across this polar desert of fallen meteorites; a place of implausible superlatives that seemed out of place.

* * *

Clouds linger just above the icebergs, sitting in the Murchison Sound, sculpted by the wind, and then a flash of color from brightly painted wooden houses—almost the first evidence of human habitation for the best part of a thousand miles. An onward helicopter flight; a pre-pubescent excitement is pumping through my veins. My face is glued to the window examining the otherworldliness beneath me, the unfeasible dimensions that come into focus. The idea of living in the Arctic tundra had been bouncing around my mind for years. And here I am, finally.

The thudding red machine lands on a flat bit of rock right in the middle of a speck on the map called Savissivik. The entire population has turned out. The post is eagerly unloaded. I am the only passenger, and the gathering soon disperses except for two men (Titken and Jens Ole) and a group of smiley children who are magnetized to the stranger. Titken is somber and stern-faced whereas Jens Ole is constantly grinning. Jens Ole, a retired hunter, takes me to the house straight away which he has warmed up for me. The hut is red; A-framed and located right in the middle of the settlement, in front of the school. The door has a polar bear proof handle which you have to push up to open. On the way to the hut, there is a brief encounter with a couple who drag jerry cans. Sighs, grins and a few words are traded. One of them says *taima* ('enough') and the conversation is gonged to an end.

Inside the hut, Jens Ole and I are assaulted by an armada of semi-anaesthetized flies. One alights on my chin. Window sills act as funeral parlors for regiments of the fallen. Dodging the flies dropping from the ceiling, I can see that the accommodation is basic, but that there are some antique pieces of furniture. There is no oven, but the coziness of the hut is instantly appealing. This is what I had hoped for. There are two very large velvet red settees placed opposite each other and a single bed in the corner of the room. There is a television with one channel. There are no subtitles: the Greenlandic words are too long to fit on the screen. The bathroom—a loo seat perched on a bucket lined with a yellow bag—is located in a narrow alcove by the entrance. The bathroom is so narrow in fact I am only able to fully stand up sideways which is going to be something of an inconvenience. I wonder how this will work. In the spare room, there is an old dilapidated piano. I cast a glance at the oil heater—the axis of my world for the year to come, I suspect—warbling in the corner.

I ask Jens about getting oil for the heater and he gives me some factual information about the settlement. Families of four live in cabins smaller than mine. Larger houses stand empty for they cannot be heated. Some people like his cousin Bodil only cook over the oil heater. Very few people have an Internet connection; the phone network might be down for several weeks during the winter storms. Power cuts are common. Nobody has running water. Savissivik got electricity in 1989. There is a small generator at the other end of the settlement. The person who manages it, Kuulut, is away hunting eider ducks. Savissivik got oil heating in the 1960s. Before that, there was coal. Before that, whale blubber. I already warm to this place, and the apparent shortcomings only add to its appeal.

Jens takes me to the tiny shop to buy some rations—oats, Lithuanian condensed milk, strawberry jam, chocolate—all bearing a January expiry date. The shop is about the size of my cabin with an annex that acts as the post office/air traffic control/weather station. The post is ordered into five boxes: Duneq, Kristiansen, Petersen, Jensen and Qujaukitsoq. Outside the shop, rows of jerry cans have been filled with oil and are waiting to be collected. It is a system based on trust. One by one, hunters filter into the shop, shaking my hand and sharing a joke with me or about me—I am not sure which. I buy some oil and Jens's brother (Vittus) writes *tuluk* ('Englishman') on the receipt. I joke with him that it should be *ittuluk* ('the old man') and there are tears of laughter. I ask about Internet access. Vittus shrugs his shoulders and laughs:

'Not working now. In winter, better when signal bounces off the ice. *Ammaqa* ('maybe').'

Shortly after midnight. The late sky awash in crimson. An easterly breeze speckles the sapphire-colored sea. Two hunters—Alutsiaq and Qulutenguaq—invite me into a wooden, A-framed house. Alutsiaq's face is long and chiseled. He has a shamanic air to him. Casualty of the cold, he is missing three fingers. Qulutenguaq has a round, florid face. His mischievous eyes dart around narrow sockets, crinkled from laughter. Between the two of them, the men have about five teeth. Piled up against the hut are rusting prams, discarded toys and bits of broken furniture. They wish to know who I am and what my purpose is here. The two men are born conspirators suspecting that I am a 'Greenpeace spy from Norway':

'Greenpeace *ajorpoq, ajorpoq* ('bad')', they say repeatedly.

I explain that I am a friend of Ane, a local priest, and that I will be staying in her cabin. I tell them I am interested in their way of life. Silent nodding, a faint chorus of *iih* ('yes') then more silence. They knew all this before. News travels fast in remote places.

But why am I really here? I suppose I want to know what life might be like if we choose another path.

* * *

Clothes and shoes are piled up in the porch. Dried narwhal skin (*nikkoq*) hangs on lines; dead arctic hares (*ukaliq*) on hooks; skinned arctic foxes (*tiriganniaq*) on gibbets. The pungent whiff of blubber and sea mammals hits me as I enter. The kitchen is crammed with dirty plates piled up on one another, a dried-up narwhal steak sits in a frying pan. Stepping over the narwhal blubber spread out on newspapers on the kitchen floor, I am channeled into the corner settee in the chaotic living room and introduced to the various ancestors whose photographs line the walls, one of whom, Qâvigarssuaq (the name means 'big eider duck'), travelled with Knud Rasmussen (the Danish-Greenlandic explorer who lived here a hundred years ago) to Alaska.

From our conversation, it is clear that the hunters do not approve of any kind of urban life (such as the local town, Qaanaaq, 120 miles from here) where in their opinion Lords of jealousy reign over spoiled futures. Equally, they do not trust Americans, Chinese or anybody that lives apart from nature *hamani* ('down there'—by which they mean the rest of the world). Whilst I chat to Qulutenguaq, Alutsiaq whacks the remains of a frozen fish with a hammer. Dinner is about to be served. We drink tea from a Thermos and chew on frozen halibut and *mattak* ('the skin and blubber of a narwhal'). Both men take turns in explaining to me why 'they must hunt' and how they cannot live without *mattak*—their only real source of Vitamin C. I don't need convincing. The actions of a couple of dozen hunters are detriment to nobody. It takes me forever to soften up the rubbery *mattak* in my mouth before I can actually swallow it. I try to conceal small chunks of it around my gums, but they have seen visitors play this game before. 'Swallow, swallow', shouts Qulutenguaq. We talk about their dog teams, and they promise to help me train a few of their puppies. I will not hunt. I am the guest here. But running my own dog team and travelling across the sea ice for the sheer sake of it is the stuff of dreams.

It has been a long day. After having met the latest members of Alutsiaq's dog team, I make my excuses. Outside, the primordial sounds that define Greenland. A demonic din. The shrieking, squealing and howling of the dogs is unceasing in the near perpetual light.

A thought after my conversation with the hunters: too much of the paraphernalia of modern life conspires to eliminate our dreams. Prisoners of monotony, so many people invest their entire lives pursuing 'careers' so they can afford to buy piles of consumer goods which nobody ever needs. They only choose to *live* once it is too late.

August 21

A few basic facts about north-west Greenland:

- Our GPS coordinates are: 76.0195° N, 65.1082° W. We are 900 miles from the North Pole
- 700 Inuit live in four different settlements that cover an area the size of Germany. There are no roads between the settlements
- the Inuit of north-west Greenland are the last Inuit to travel exclusively by dog sled over the sea ice
- the sea ice normally forms in November/December time and disappears in June/July
- Greenland is 80 percent ice. The Greenland ice sheet covers the whole of Greenland from east to west, and lay a few miles from my hut. In places, the ice sheet is about two miles thick
- my hut is located one hundred and fifty feet from the shore
- the sun goes down for the last time on October 24 and only reappears above the horizon on February 17. From mid-April to mid-August, there is twenty-four-hour sunlight
- up until 1818 when Sir John Ross came to north-west Greenland in search of the Northwest Passage, this subgroup of the Inuit believed they were the only people on the planet

* * *

I awake to the sound of dogs chained outside the hunters' homes. Anguished cries, tormented pleas and frenzied heckles. Some kind of massacre appears to be unfolding. Dogs are taking up their positions in the hierarchy of the pack.

I spend the morning getting settled into the hut. The cabin looks out onto the bay; retreating glaciers and the twinkling ice sheet in the far distance. The air is so clean and dry you can see for nearly a hundred miles. Through the window, beauty reveals itself. No book has prepared me for this. My soul is spread open with satisfaction. Reality is not always disappointing after all. I am almost upside-down with emotion. Is this my life or the one knit together through dreams? Glossy icebergs straggle the mottled water, occasionally exploding in the summer sunshine. Billeted in the High Arctic for the summer, cheerful snow buntings chatter by the window. Sunlight floods the hut. I sit for a while and watch the arc of the sun. Soon, the window will be frosted over and ice will invade everything.

Outside, it is four degrees but feels much warmer. Down by the shore, beyond the pastel-colored houses, rhombus-shaped lumps of ice measuring seven feet across are parked on the beach. Ice; beach: a curious juxtaposition. The icebergs I could see from my window are the size of small English villages. I walk the length of the beach. Empty shipping containers, a dead harp seal, its wishbone shaped markings stained in blood. Further along still, an up-turned sled, rusty pram wheels and then a decaying dead dog.

Soon I meet a hunter called Ole. Ole is proud, smiling, an eater of arctic foxes. His large family is scattered across north-west Greenland. He cuts up seal, his flabby girth reaching well beyond the confines of his T-shirt. He tosses lumps of bloody chocolate-colored flesh into a wheelbarrow to feed his eighteen dogs. The hunters keep their meat in blood-stained wooden chests called *nequahivik*. In the severe cold, dogs may be fed two or three times a day. In the summer, every other day if they are lucky. Ole's wife keeps puppies at bay. They are desperate for any scraps they can get. Soon, it will be feeding time and the pandemonium is just beginning. The barking and howling escalates into an infernal crescendo that echoes around the settlement. The chained dogs are hysterical, launching themselves at Ole and practically strangling themselves.

I watch the feeding ceremony with intense interest for I hope to look after some dogs of my own. Ole casts aside the seal skins. 'No market for these anymore', he huffs and puffs. Lumps of seal meat are tossed to the ravenous dogs, one-by-one. Pomegranate red blood drips from their jaws.

Ole and his wife invite me into their home where children run along the spine of the roof. A young boy, Felix, throws darts at dead seals lined up on the kitchen floor. A violent, expletive-filled Hollywoodian film aimed at global-villagers plays from the living room where teenagers dressed in baggy jeans slurp on seal soup. I am devastated.

In the relative calm of the kitchen, no questions are asked of me. We chew on dried narwhal skin. It looks like burnt wood. Ole guzzles down two liters (half a gallon) of coffee. 'Me, originally from Moriusaq', he proclaims and reaches for the Greenland telephone book pointing to the entry with a population of one:

> **Moriusaq**
> Jakob Hensen, 97 46 78
> Nukissiorfiit (electricity), 94 23 81
> Sundhedsvæsenet Moriusaq (health center), 91 26 78

He talks about procreation and hunting. The two seem to go hand-in-hand somehow. Ole, a local Othello according to Titken, is angry that they are allowed to kill so few narwhals this year, but still cannot quite dispense with the grin. *Ajorpoq, ajor* . . . The conversation peters out into a series of misunderstandings, *non sequiturs* and high-spirited laughter. I apologize for destroying the Greenlandic grammar. He turns the radio on and I take it as my cue to leave. I return to the silence and solace of my cabin, not wishing to be away from the views from the window for too long. I immerse myself in the rhythms that come naturally. The heater is spluttering and snoring in the corner. The sky has turned into Chagall's *Ceiling of the Paris Opera*. All Byzantine coppers and golds. Lens-shaped clouds the shape of UFOs take my thoughts away to a universe of numinous experiences. The evening sunlight shines on these sooty puffs, burning them shades of pink, red and orange. I reach for my notebook and jot down poetry and aphorisms which provide scaffolding for memory. Who needs a television when you have a window onto the High Arctic?

Dead seals lined up in Ole's cabin

August 22

It's morning. Four o'clock. The sun has already seized the hut. The brightness is raw and unrelenting. I can hear noise.

The Captain Khlebnikov is in the bay. This is a Russian ice-breaker that takes well-heeled tourists to remote parts of the Arctic. It is dwarfed by the adjacent cathedral of ice. The boat goes north from here, up the Nares Strait and into the pack ice. The conspicuous tourists appear with their bright yellow coats, carrying cameras with vast telephoto zoom lenses. The atmosphere changes for the few hours they are here. I feel as if I am in a zoo. They photograph everything. They wander around the settlement commenting on the poverty and 'the primitive way of life'. They look at me as if I am a piece of driftwood washed up on the shore, and then get back on their boat to the 'furthest North'.

On the ship there are helicopters which fly small groups over the ice sheet immediately behind me. Everything is accessible now. If you pay somebody enough money, they will take you to the North Pole or push you up Mount Everest. The noise of machines lasts all afternoon. They are attacking the silence. I came here to escape herd behavior, machines and noise and now it has come to me. I just want these people to leave. The slightest intrusion from my 'fellow men', and I feel some kind of panic. The castaway syndrome has set in already.

August 23

The noise of men and machines recedes and then disappears. It is replaced with the calls of the wild, the howling of dogs, the moaning of the puppy orphans. No longer a din, I already love the melody for it reveals somehow the *genius loci* of the place.

Not a cloud in the sky. The ark of the sun has never interested me before. Now, it is a source of fascination. It rises from behind the cliffs on the cemetery side of the settlement and falls towards Northumberland Island. I put my watch in the drawer. I won't be needing it. This year, I will not be a prisoner to time. By looking at where the sun is in the sky, I have a reasonably accurate idea of what the time is and what I should have achieved: collecting the water, making the porridge, the piles of washing up, slopping the oil into the tank, the Shelley poem. When it is dark all the time, I will need instead the stars to guide me.

My duties done; a toothless grinning man enters the hut. A stranger. He removes his shoes and makes a beeline for the sofa as if this were a familiar routine. He sits in silence, peering out the window focusing his gaze on what matters most to him. This is the way of those who live alone. I introduce myself. He grins. *iih, iih*. I ask him if he would like coffee. He raises his eyebrows ('yes'). I prepare the Thermos and buy myself a bit of time to formulate the next question. A few more questions. He sighs, raises his eyebrows and occasionally screws up the skin at the bridge of his nose ('no'). There is no need to keep a conversation going here. After about half an hour, I have ascertained one thing. His name is Qaaqutsiaq Qujaukitsoq. Then, he is gone.

* * *

A few more basic facts about Savissivik:

- out of the population of forty, there are eighteen bachelors
- there are approximately seventy dogs
- one tiny school (five children); one tiny church (normal attendance varies between three and six people); one tiny store where basic supplies can be bought
- a supply ship visits the settlement every summer from Denmark
- houses are heated using oil. Oil comes on a ship once a year
- there is no waste drainage
- Most hunters live in more or less the same houses, just painted a different color. Some own dogs, some do not. Some are better hunters than others, but no fuss is made about that. Everybody shares what they have, children (and sometimes wives) included.

August 24

Morning has started. A dog is screaming for its life. My snow bunting squats on the window sill. A raven rearranges the tools on my balcony, planning I am not sure what. Then, he stops and studies the lay of the land. Through my binoculars, I watch a pod of beluga whales pass through the bay, escorted by a small colony of obstreperous arctic terns. Lonely castles of ice turning blue sit like moored ships. Later, a narrow channel of cloud runs right across the horizon dividing the panorama into wafers of different shades of blue and wishy-washy grey—the seascape and landscape becomes one. Only once the view from my window is smudged, do I look for some company.

Puppies follow me everywhere I go, wrestling with the laces on my boots. Soon, they become my shadows. A team of dogs are fighting over a baby seal. Scarlet gashes. Inflamed with jealousy, one scruffy, wolf-like dog (I'll call him Lenin) attempts to reinstate the hierarchy. Adjacent dog teams are blood-thirsty spectators. Eventually, a hunter appears and hits Lenin's culprits with a steel bar. Some kind of order is maintained.

I have only been here a week and am already anthropomorphizing dogs and thinking about naturalizing men.

* * *

Aleqatsiaq—a man entangled in nature—smokes a pipe on the balcony. I met him at the shop on my first day. He waves at me and says abruptly *kaffi* ('coffee'). *kiak, kiak* ('hot'), he moans as I project myself into his hut. A dead narwhal hangs suspended from the kitchen ceiling (travel can become dull when you just see ordinary people doing ordinary things. Thankfully, that seldom seems to happen here). *Ooh*, I say. I want to try the meat and my request is met with an imperative, *neri, neri* 'eat'. The meat is rubbery, squid-like. 'The killer whales in the bay are scaring off the narwhals. Hunting difficult', he complains. *Ajor*.

A pause and then Aleqatsiaq starts talking about a *kivitok*. He has seen one down on the beach, half-man, half-beast. Serious nodding. No laughter. Animals are not just the masks of the human image here; animals and humans share features and personalities. Intermingling mythologies. *Kivitok* are outcasts who cannot integrate into the community. It is believed that they have special powers. Loners, they sometimes walk out onto the ice cap where they die.

A grandmother sits on the sofa, rolling cigarettes. A young child who addresses me as 'grandfather' skips around the living room. I sit on the sofa and embarrassingly break one of the springs. The grandmother does not look impressed. I apologize as much as I am able to in a *mêlée* of broken grammar. She is sowing a pair of braces into a new pair of *nannut* ('polar bear fur trousers') for her son, Aleqatsiaq:

'The hair of the old polar bears is thicker and warmer than the thin and brittle hairs of the younger animals', she says.

We talk about the local diet. Polar bear *mammaqtoq* ('delicious'), they agree. The meat is eaten on its own and has to be boiled for twenty-four hours to avoid trichinosis:

'Only Ole eats fox', Aleqatsiaq laughs. 'He got used to it as a child. Probably alright with salt', he adds. 'You can eat everything here', he says, but *tugto nerukka* ('caribou intestines') I do not recommend':

Kiak, kiak ('hot'), he repeats, pushing the window wide open. It has never been this hot in the Arctic. They say it each year apparently. The water is pouring off the ice sheet. A way of life will disappear once the sea ice has melted completely:

'There are too many *kad'luna* ('white Europeans or non-Inuit'). Our culture will die when we are no longer able to travel by dog sled over the sea ice', he holds forth.

I lower my head, acknowledging our guilt perhaps. He is right of course, even if the moral world cannot be dissected as neatly as the physical world.

The conversation palls. Silence is again sovereign. Aleqatsiaq says he can hear the walruses talking by the shore. [I am reminded that there is still so much we don't understand. We all act as if we have a *total* view, but there is so much we are ignorant of]. He moved here for the silence, the simple life. He is quiet and respectful, discrete and unimposing. More silence. These coffee-talks seem to have something of the restrictions of an art-form that I have not yet mastered, and the transience of mortality.

* * *

My thoughts upon these visits: the age of living *in* nature has past, but these hunters still possess the keys to bring its magic to life. Free from the constraints of work, they still know the path to freedom. They possess time. Time does not rule them. It is only when man loses the memory of such a life lived in equilibrium that he no longer recognizes the source of his unhappiness.

August 25

I sit in the hut eating Swedish chocolate which I dedicate to the memory of past indulgence. I am overcome by the desire to go walking on the ice sheet, across all those trillions of gallons of frozen water. There are no mammals there. Not even polar bears which are found foraging instead down by the shore. And so, I am busy consulting the maps of my mind. Am I seeking some kind of void? All those blank spaces seem to have an uncanny appeal. The urbanized man seeks *terra incognita*, it seems. These urges cannot be explained by those who live *in* nature. For years, that two-mile-thick sheet of ice that meringues the whole of Greenland has tempted me. There are at least some thoughts that don't have to be tamed, and I reckon this magical place has provided the means.

I make preparations, quickly filling a rucksack with base layers, gloves, tea, chocolate etc. I walk behind the settlement with a line of cappuccino-colored puppies in tow. Soon, we are out amongst a desert of rubble and stone; the dogs finally have the sense to give up their pursuit. The sky above my head is bronze, the ground beneath me iron. Deuteronomic scenes. Devoted to freedom, I am walking in the spoils of meteorites across mountains of scree, dumped there by something greater than man. I consume the rocky mounds. There is the occasional cairn (*hakamattaq*)—navigation markers and a reminder that this is *only terra incognita* for the western mind. After a couple of hours of sweat and toil, I see the Greenland ice sheet before me. At first, I am not sure if the layer of white on this cake of igneous rock—spliced and incongruous—stretching right across the horizon is cloud or ice. Distance and scale are almost impossible to judge in the Arctic. The rays of the sun bend and twist—a polar Fata Morgana; distorting and confusing. A place without fixed definition; a vast expanse of lunar infinitude. The consciousness of space fills my being. I pause—silence, but for the thump of my own heart. I look around me. This landscape is incapable of hiding anything. I continue to labor over rock, lichen-covered boulders with painted faces and stone for over two hours, but have still not reached the elusive ice. The Arctic desert landscape has conspired to fool me. What seems so tantalizingly close is in fact many miles away. The axiom of the desert.

The air becomes brittle. I return to the hut, crossing my repository of freedom. Make tea. Read Auden. Indulge in the fault lines of language.

* * *

My thoughts on life today: the purpose of life is surely gratitude. To be grateful for the beauty of this world and to be thankful for any transcendences that flash us the totality of life.

August 26

Woke up early. Patriarch Lenin was making hell of a racket. The thin curtains punctuated by sunshine. I cook porridge over the choking oil heater. The routines that govern my life here are already becoming familiar. I almost revel in their anti-modernity. Once I have collected the water, I am off again exploring landscapes at the edge of infinity. I wish to see everything I can before the darkness sets in. I follow the shoreline, walking beyond the settlement. Regiments of flies overhead. I soon come across a tatty looking dog. Far larger than any other I have seen. He watches me approach what he clearly regards as his territory. This must be a lead (sled) dog. He sits there proudly like Cerberus guarding the entrance to Hades. Not able to see if he is chained, I am unsure whether I should proceed. I risk it and the creature stares at me, watching every step of mine, but thankfully stays put.

Around the corner of the bay awaits a sandy beach. With the exception of the odd fragment of a whale carcass, the beach is empty. The midday sun is hot on my cheeks, and yet I am almost at the top of the world. The beach becomes pebbly, and then littered with gargantuan boulders painted with lichen. Each face and facet a surprise. A geologist's dream—but I can't get excited about rocks. The way becomes difficult to negotiate for a while. I come across another dead dog. The beach is beginning to feel like a deserted battlefield, scattered with the corpses of the losers. Who was the despot this time?

At four in the afternoon, the 'ship' (Royal Arctic Line) arrives. This is the ship that comes with all manner of supplies in the summer of each year, weather permitting. Unloading will begin at high tide. Previously, the bay was sometimes frozen over even in September and so there was no guarantee that the supplies could reach the settlement. I think I would have liked to have lived in those times, to be forced occasionally to live *in* nature and survive *from* it.

Even here, we are connected now to that 'world of perplexity'. Dependent in some way on consumer goods. I thought I was privileged to escape that life, to traverse the equator of my consciousness. Baudelaire dreamed of escaping the reminders of 'the everyday' and I seem to harbor the same desire. I thought I could sit up here for a year and just be an observer to humanity having gone astray. I can already see that will not be the case. We are connected too to the global marketplace. It is true that we live here in tiny huts with no running water, but Amazon delivers.

August 27

I go to the church service in the morning. The atmosphere is stony and solemn. Not an ounce of humor. I am thankful for Jens Ole's grin parked at the end of the pew. He has one eye on the seal skins that flutter in the breeze outside his house. That is his world. The church bells ring, but not for any amount of time. A duty is being fulfilled. The church is tiny. Fascinatingly so. There are four of us; the faces all by now familiar. A lay-preacher (*ajoqi*) and a congregation of three. I sit on one of the pews at the back. I watch the shadows slide down the walls. I am little more than a curious spectator here. There is no sermon or eucharist, just readings about sin and sinfulness. The decalogue is read in a muttering staccato. The lay-preacher stumbles over the syllables of words that stretch confidently across the page; an assault of popping sounds. The service is a reminder of our collective guilt and transgressions. Surely, religion should be more than a penal code. Sin is only valuable when it is liberating and leads to God.

After the service, I get talking to an elderly man wearing black wrap-around glasses. His frame is wiry, all ligament and sinew; his face etched with stories; his voice choked with memories. He lives in a tiny blue cabin. I try and keep the conversation going, hoping that he will invite me in. I am in luck.

A man at the end of his life is living in one room about eleven feet square: chaotic, cigarette-stained, overheated. There seems to be no bathroom. His hut is the realm of absolute concepts: subsistence, family and goodwill. Life here is stripped to its basics. On the wall hang scores of family photographs, memorabilia from polar journeys, faded pictures of Christ, lop-sided crucifixes, a birth certificate and hunting diplomas. A factory for memories. I overlook the dirt and faint smell of tobacco. He lights a candle bidding me welcome and pours me a cup of tea from a grubby Thermos, rolls a cigarette and tells me repeatedly that I, a *takornaqtoq* ('stranger', literally 'somebody not seen before'), am welcome in his home. Taateraaq, a man of a gentler century, has almost no material possessions. No Girardian mimetic desire here. No 'wanting what we want because other people want it'. He points to the smeared photos of his five sons on the wall. Images of joy, happiness and cheer repeat themselves. He has his boys, the stories of a life lived on the polar ice and is totally content. He sacrificed an eye to one of those journeys.

In the ice is enshrined a sacred life, an ancestral spirit that will be lost once it has melted away. This is what I take away from Taateraaq's words. Taateraaq shares his soul and spirit with the animals and birds. Even his name means 'kittiwake'. He lives from the sea ice and the land: walrus, seal, polar bear, hare, birds etc. The animal world taught him how to live close to the earth, and together they share the pulse of the visible and the invisible. He has never taken from the capital of the Earth, but is instead somehow a part of it. For him at least, there was never much of a grocer to complicate the spiritual issue. I bask in the sensation of his modesty.

Mortgaged to nothing, he lives in kinship with fellow-creatures. He is free, not enmeshed in some huge mechanism estranged from the mysteries of wilderness. For the rest of us, it is difficult to escape the machine. He tells me the story of the first time he saw an airplane fly overhead. He had no idea what this could be. These machines were seen as giant insects. Local people used to call helicopters *niviuvaarhut* ('midges'). Now, we are more surprised to see a flock of cranes fly overhead than human beings in flight.

Later on in the day, down by the shore and twined in puppies who do not care for my abstraction, I see the catechist with a rifle slung over her shoulder (all the guns are kept in wooden chests down by the beach). Dispensed with the ruff, she makes her way towards a small motor-boat. The catechist is remarkably shy, struggling to make eye contact and clearly not entirely comfortable speaking to a foreigner. I ascertain that she is off to shoot ducks and then she is gone.

Note: the mistaken metaphysics of long since dead Western thinkers has left scars that only now are being tended to. I am thinking of the alienation between us humans and the planet we live on, the alienation between us and animal species, the alienation between our thoughts and our bodies. Only now, after centuries of misguided philosophy, are we finding a new humility. We should have listened to Blake:

> "How do you know but every bird
> that cuts the airy way
> Is an immense world of delight,
> closed by your senses five?"
>
> (*The Marriage of Heaven and Hell*)

August 28

The first dew. Mist cloaks the houses. Spiderwebs suspended under moisture. Purple saxifrage—smiling embellishments of the tundra—gleams in the moss. I pick their leaves and stems and brew some herbal tea in the hut. I read Eliot's *Gerontion*. It may not be 'depraved May' and there is no dogwood, chestnut or flowering judas here, but I too could be 'drunk among whispers'.

Mist lingers over the sea for much of the afternoon. The sea looks grey. I sit waiting for the curtain of mist to tear, but nothing. With a flick of the switch, the world has turned from color to the black and white of an Ansel Adams photograph. Gone are the shiny, vivid hues of the endless Arctic summer. The cliffs to the east with their concertinaed folds have been radiating red and purple in the late evening summer sun, but through the *pujoq* ('mist, smoke') now seem uncompromisingly white.

A quick picture of the place: down at the shore, some hunters pore over the latest catch. We exchange a few words. I scrape around amongst my private aural histories for the feel of the language. The rhythm of conversations is slow here. People pause over the words. Nobody interrupts you mid-flow. Nobody feels the need to make conversation. I like that. Those not hunting sit in their overheated homes glued to Danish cartoons and Mr. Bean. Hip-hop music blares from a blue cabin behind the water tanks. Teenagers wave as I saunter past. A man stripped to his waist smiles from a cracked kitchen window. An ageing hunter pushes a wheelbarrow full of fresh blubber up a steep, sandy slope, leaving a trace of seal blood behind him.

In the evening, the sky clears and the temperature drops. The oil heater splutters and burps before cutting out completely. Once it has cooled down, I start to take it apart and soon identify a blockage in the cable connecting the heater and the tank. I retire to my sleeping bag having settled a few scores with the heater. I read Byron who, as always, is busy romanticizing himself from the word 'go', and fall asleep to the comforting contrast between the warmth of the heater and the frigid outside:

> Those flaxen locks, those eyes of blue,
> Bright as thy mother's in their hue,
> Those rosy lips, whose dimples play,
> And smile to steal the heart away,
> Recall a scene of former joy,
> And touch thy father's heart, my Boy
>
> (*Fugitive Pieces*)

August 29

In the fatigue of the early morning, I wake up to the creaking of a door. 'Stiffi', 'Stiffi'. A stranger stands in my cabin. Blurry-eyed, I try to find out who he is and what he wants. His name is Qi'ddut, the husband of Ane who owns the house. He has come for the rent, but how he got here I have no idea for he does not live in the community. I fumble around looking for my cash. I give him the money. He goes outside, fetches a set of ladders, opens up the ceiling trap to the loft and brings down a small box which acts as a safe. He then puts the money back into the loft, knowing full well that I could take it back down if I so fancied. His banking facility is my house. *Qujan* ('thank you') he says and then is gone.

I tend to my chores: washing, shaving, eating breakfast and making notes—all are symbolic and meditative acts of celebration. As Somerset Maugham once said: "there lies a philosophy in each shave." In the afternoon, I abandon myself to the sunshine that streams into the hut. I listen to the dogs howl and screech. Lenin's compatriots are foundering in blood. Under a blaze of pale, pink light I read Dylan Thomas. Like a pebble thrown into a pond, his private metaphysical vision spreads in all directions:

> Bell believe or fear that the rustic shade or spell
> Shall harrow and snow the blood while you ride wide and near,
> For who unmanningly haunts the mountain ravened eaves
> Or skulks in the dell moon but moonshine echoing clear
> From the starred well?
> A hill touches an angel. Out of a saint's cell
> The nightbird lauds through nunneries and domes of leaves
>
> (*In Country Sleep*)

Byron thought Wordsworth's poetry was namby-pamby. One could hardly level the same accusation at Thomas.

The evening sky is aburst with red fire, the color of authority and insurrection. I sit in my candle-lit hut, gazing out the window. Even when I am in my sleeping bag and half-asleep, it is difficult not to peak through the curtains at the insane drama spread across the sky. The screaming continues. In the light of Lenin's dictatorship of terror, some of the dogs seem to be consumed by a demonic agitation. Eventually, I drop off.

30 August

A night of nocturnal wrestling with dreams of helicopter travel. Snow geese fly over the settlement. I feel as if they must symbolize something. In small flocks, they fly effortlessly, but at great speeds, across the crystal blue sky. Great skuas squabble on icebergs washed up on the beach. The sun is beginning to sit lower in the sky, its metaphysical hues spread across a sweep of colors.

Today begins the dog training. I speak to Alutsiaq. He sends me up the slope, past blazes of saxifrage garnished with beads of dew, to meet

Naimmanitsoq who will help me. On the floor of his kitchen sits a bowl with a seal skin in water, a bucket with a musk ox head and another small bowl that contains reindeer meat. Naimmanitsoq is in his early fifties. An active hunter whose strength is drawn from his animal volition. A thick-set man with an amiable face. Unless talking about hunting, he speaks few words. He is repairing his *nannut* and asks me to come back later. He hands me a gift of some fresh musk ox meat. *Mamaqtoq* ('delicious'), he says and smiles with his eyes.

At the hut, I fry the steak over the oil heater. Delicious with some mustard which I managed to find in the replenished store. I spend the afternoon training with the extremely long whip (*iparautaq*) which Naimmanitsoq unfurls with perfect precision. I walk around the settlement with one dog that he has lent me. *Nipi* ('voice; sound') is three years old. A strong, but obedient female. I try to control her movements using the local commands and by cracking the whip occasionally alongside me. *Harru, harru* ('left'), *atsuk, atsuk* ('right') and *aulaitsit* ('sit still'). The hunters are bemused by an outsider training a dog Inuit style. They weren't expecting this. I circumambulate the huts until the shadows fall. I am happy to bond with *Nipi*. Slowly, I will begin to train them in pairs and then perhaps in larger groups.

August 31

In the morning, three young children with names of dead ancestors creep into the hut, giggling. They have been following me around like little conspicuous spies. This cohort of smiles scrutinize the blond hairs on my arms which they are fascinated by. I show them how to take a photo with my camera. They look at the images; surprised to see themselves. We do some drawings and they write down some words for me in the local language. As words and idioms become jumbled in my mind, I feel like the most useful word would be the equivalent of *thingamajig*. I am intrigued by the term *akutsihuq* which refers to somebody who has several children of alternate sex: son-daughter-son-daughter. After a few card games and some psychedelic-colored fruit juice, they announce that they must go. And off they skip down the steps to join the puppies chasing their tails outside my hut.

Thought in the afternoon: it is extraordinary that the Inuit were abducted by the polar explorer Robert Peary and taken to New York where they became exotica; objects of curiosity. Most of them died of tuberculosis. A young boy called Minik was amongst them. His father died; his skeleton became a museum exhibit. That was just over one hundred years ago. The horrors of that age. Who wants to be put on display in a cabinet?

On children: in the West (and perhaps elsewhere), so many wish to have children for selfish reasons. They want part of themselves, their reputation and name to continue into the future. When they find out their children share little of them because they are the product of an overwhelming consumerist and technological age, they are, I suspect, a little disappointed.

Children playing in the author's cabin

September 1

A small flurry overnight has now turned into heavier snowfall. Suddenly and without warning, the seasons have changed. This rhythmically recurrent order in life—the influence of the seasons—is much diminished in societies that we call 'civilized'. We spend our days in cocoon like machines insulated from the changing chords of nature. The rhythm of my life here is already more or less determined purely by natural phenomena and not human incursions. This is what I was looking for.

The outside temperature has dropped to minus ten degrees Celsius (14 degrees Fahrenheit) and is forecast to fall as low as minus twenty-five degrees next week. Still, this does not stop Ibbi coming to visit in the evening wearing just a thin blue jacket with a T-shirt underneath. Poring over maps, Ibbi, a part-time hunter, shows me where he has been hunting in the Bowdoin fjord (a nearby fjord named after Robert Peary's *alma mater*—Bowdoin College). We share a few beers and after a great exchange of genealogical information, we soon find firm ground for our commonality. He describes himself as a *heqaijuk* ('a lazy, apathetic person'). I am grateful for his company. He disappears occasionally for a smoke on the balcony and is soon lost in tobacco. It occurs to me that smokers have more to say than non-smokers. Then, he returns:

Ibbi: 'I shot a *kivihoqtoq* ('seal with little blubber that sinks easily'), but it sunk.'

Much movement of eyebrows.

Ibbi: 'this map is out of date. This glacier [pointing to Josephine Peary Island, (Qeqertaarrhhuharhuaq)] is no longer connected to the Tracy Gletscher (Qeqertaarrhhuharhup hermia), but is an island unto itself. This change has happened over the last ten years.'

Ibbi looks sad. He shrugs his shoulders.

Ibbi: 'the calving of icebergs in this area has become a problem, affecting narwhal hunting in the summer.'

Many of the glaciers in the region have receded so rapidly that they are no longer productive: the once icy fingers that reached the fjord, are now greying stubs. Uncertainty about tomorrow.

Ibbi: 'hunters used to travel over the Greenland ice sheet by going up the nearby Politiken Glacier. But the glacier no longer reaches the sea and the dogs will not pull sleds over rocks.'

Ibbi: 'And then the snow. There is far less snow than there used to be. And so, the dark period seems darker than it did previously. We know it is much warmer now for the dogs' frozen breath is not so dense in the winter.'

The human economy is forcing us to exploit the planet. If economic growth is the only parameter of 'success', how will we ever slow down and preserve the remaining wild places? How could a degrowther who would have to oversee a contraction in the economy ever be elected to Office? Old thinking prevails.

September 2

My mind is already working in a different way. The intellectual gymnastics of Cambridge are a distant memory. Intellectuality has been replaced by a spiritual aura. The sounds of nature colonize my hearing now, and so I have rediscovered the privilege of being in touch with my acoustic intuitions. I speak slower, walk slower and think largely in poetic aphorisms. My eyes no longer greedily devour the words of books. I spend my time molding my thoughts about humanity. I can feel the shape of the words in my mouth, the weight of the syllables on my tongue. Like all language learners, I have become a quasi-other.

There is nothing to be gained from being quick-witted here. No audience for *doubles entendres*. Mental agility and repartee have been replaced with sighs, groans and the slow polishing of Spenserian stanzas. I wonder if I will forget entirely the art of conversation, whether I will become less worldly, seeking out silence in the corners of the Earth?

The crunch of fresh snow. I walk with *Nipi*, my lead dog, my only dog, practicing commands until shadows take over the settlement. Infatuated eyes. She smiles at me. The currency of her affections. The other dogs are sleeping. Silence and peace everywhere. This solitude and uncluttered landscape fill me with an inner freedom. Who I can be and what I can achieve is now generated solely by me. A simple glance at the sky—crimson jungle—and I know I have made the right decision coming here. This life, this view. Just being here is enough. The days might not be eventful, but I am collecting oil for the lamp of memories that will burn forever. These memories have a transcendent power for they earn me interest on time.

So, I am happy watching the flame flicker in the sky, the distant cliffs shine in the twilight. Beauty might not save the world as Dostoevsky once said, but it saves my evening. Sitting amidst all this perfection, I read aloud MacNeice's *Autumn Journal* in my sleeping bag wondering whether the frost out there will "kill the germs of laissez-faire." He was writing a journal too, and his archive of public events still echoes in this decade of uncertainty. All the pressures of his time, the 1930s, were building up in some kind of socio-economic combustion chamber. He stockpiles these concerns and then releases them with personal inscription and propulsive *élan*. Occasionally, I pause to reflect on those who seem comparatively speaking at least entwined in chaotic turmoil *hamani* ('down there'), that world outside. And then I fall asleep to the baying of the saints and sinners, happy in the belief that I enjoy somehow that privileged vantage point of the poet.

Thought on silence: the silence of nature provides some kind of spiritual allure. It is not just a refuge from human noise, but if it can be felt at a physical level so that that it reverberates through your body, it can be transcendental and ennoble the mind.

September 3

A difficult day. The howls of the dogs echo around the settlement on this frozen morning. The colorful wooden houses stand like beacons of light in an otherwise confused, indistinct landscape. Tilting and bobbing lumps of ice freckle the dove-grey sea which merges with the blurred, off-white slopes leading down to the shore. The wind has dropped, the temperature has risen and the snow falls in graceful flakes. The children that came to visit me are sledding down the snowy slopes of the tracks. 'Stiffi', 'Stiffi', they shout and wave.

I am off to visit *Nipi* when I hear a high-pitched screaming sound from behind me. I turn round to see an adult male dog about twenty feet away mauling a puppy. The adult tosses the puppy to the snowy ground and runs off. It is all over in seconds. I am just a hopeless observer. The dog has broken both of the puppy's hind legs. He is in terrible pain, tries with all of his might to walk, but at best can just drag his two disabled legs along behind him before collapsing in the snow. He is yelping and whimpering with the pain. He keeps trying to stand up, but just falls pathetically to the ground. His eyes bulge with fear and confusion.

I summon a hunter walking past. He is not interested in helping. Another hunter sorting out blubber on racks behind me tells me to kill it and points to a boulder, suggesting I should club it over the head. The kindest thing is to arrange for somebody to shoot the poor thing as soon as possible. The nearest veterinarian is a thousand miles away. I speak to Frederik in the stilted, green house just above the shoreline. This is nothing unusual for him, but I can see that he appreciates the wretchedness of it all:

'Go and find Dennis', he says.

Dennis is the 'dog killer' responsible for shooting adult dogs that break their leads. Frederik picks up the dog and puts him down by the boat with his siblings and the mother.

Dennis comes down straight away. We make a beeline for the upturned boat, but I cannot see him anywhere. He is not able to walk and it is not possible that he could have gone more than a few feet. And then,

Dennis points to the sheltered area by the heel of the upturned boat. Blood-soaked. The mother sits upright, looking away, aloof and enigmatic, but acting as a guard over the grisly proceedings beside her. A sacrificial offering. The siblings to the puppy are eating him alive, ripping off lumps of his flesh. Saturn devouring his son. Almost. We thought we had left him in the security of his mother's care, but the dogs could only see it as an offering of food. It is too late to do anything. The meal is almost over. I am overcome with shock. Dennis shrugs his shoulders. 'They must have been hungry' he says and makes off back up the slope.

* * *

Thoughts this evening: in my world, we are now so insulated from death, butchery, blood and guts. Here, it is part of daily life. A couple of generations of peace and urbanites have almost forgotten about the borders of brutality. We are perhaps shocked by these things because we know that humans too can be raw material for one another.

September 4

The sun caresses the window sill. I watch the light move across the floor of the hut. It must be nearly eight o'clock in the morning. I doze for an hour or two. A gang of snow buntings hold dawn in check. On the balcony, they converse noisily trying to entice me out of my sleeping bag. I only manage it after Alutsiaq walks into the house to collect the waste bag from the loo. It must be Tuesday. *Ajor, ajor* . . . he shouts when he sees me still in bed.

It is two degrees. Clear skies and hours of Arctic sunshine. The subject of the weather has taken on a new dimension. No longer small-talk. Weather is now all-important, shaping my every day here. Every moment. Previously, this hut has only been lived in in the summer, and the weather could determine if I can stick out the winter or not. There is nowhere else to stay.

Young children still run around with few clothes on, often not much more than a jumper. Lords of their climate. I listen to their didactic voices, calling and shouting. One of them rushes up to me and introduces herself. 'I am Ane-Sofie. You like Greenland? Can I come and visit one day?' And, then her cheery voice is gone.

A fresh breeze blows through the hut as I air it. The television speaks of the dangers of forgetting to ventilate the house. Apparently, 15 percent of school children in Greenland have tuberculosis. In my mind, I am already linking the two.

Now, that it gets dark at night, there is more of a range in temperature between night and day—not that this will last long. As the outside temperature drops, I can see how the thermometer could become a source of obsession. The lowest temperature in the house has been nine degrees Celsius. Even with quite a few layers on, it is unpleasant sitting around in the hut at that temperature. The optimal temperature inside is about eighteen degrees, but maintaining the oil heater at that temperature is no easy task. If the oil heater goes out, it takes a good hour to warm the hut.

Today, I read Keats (died of tuberculosis aged twenty-five) and other emblems of intellectual humanity. Beauty is truth; truth is beauty. I too am seeking the truth, but I am not going to claim it. I turn the pages silently, pausing now and again to watch towers of ice crash into the sea. I feel his dissatisfaction with the state of the world, but the Arctic is not a fanciful escape from the cruelty he bemoans. People die unexpectedly, animals savage one another and there are no verdant landscapes:

> When old age shall this generation waste,
> Thou shalt remain, in midst of other woe
> Than ours, a friend to man, to whom thou say'st,
> "Beauty is truth, truth beauty,—that is all
> Ye know on earth, and all ye need to know"
>
> (*Ode on a Grecian Urn*)

In addition to his lyricism, love of nature and distrust of rational forces, Keats embraced what he called 'negative capability'—the idea that we should suspend judgement about something in order to learn about it. In the hyper-accelerated world of digital social media, 'negative capability' has been pushed to the sidelines but is needed now more than ever.

As well as being here to try living another kind of life, I am also here to document the words of this fantastically complex language, to listen to the *logos* and the voices bouncing around in my acoustic memory that seem to have a life of their own. These exercises can become dramaturgical. It is like pausing at the top of a hill to hear the singular sound of the wind. The life-enhancing thoughts that can stem from such experiences far outweigh the benefits of any sterile collection of words.

I fall asleep to thoughts of ostentation in Cambridge and the shadows trespassing on my sleeping bag. A year in a sleeping bag. Can you imagine? Can there be anything less ostentatious?

Some new words: *anaqtaqtoq* 'the shit collector'; *e'qqarhautainaq* 'abstract' (literally it means 'only the thinking')

September 7

My birthday. Hurrah! To celebrate, I organize a *kaffimik* ('a party held on special occasions to mark birthdays and other events') in my hut. Ibbi, Jens Ole, Alutsiaq and the children soon join me. They remove their shoes, wish me 'happy birthday' (*pi'dduarit*) in a rather formal manner and offer small gifts of chocolates. Birthdays are taken seriously here. Birth is your link to your forebears. My high-spirited little friends race around the cabin, and this is before they have located the psychedelic juice. The other invitees are all playing Bingo in the community hall.

I light some candles and blow up some balloons that were left on the piano. Prepare lashings of coffee in multiple Thermos (living in an Arctic hut is only ever one step away from the camping experience) and put out cakes, buttered currant buns and small bite-size portions of reindeer meat on the table. My background music: set to rotating scenes of snowy landscapes, a fantastically amateurish choir singing on the one Greenlandic television channel. His eyes lit up with cordiality, Ibbi is telling jokes and wants to know why I do not have any miniature Greenlandic flags on the table. As usual, the conversation is studded with bursts of laughter. Many years ago, the Greenlanders decided they must have their own flag. There was a national competition to design the flag. The winner simply copied the flag of a Danish rowing club. Several years passed before the penny dropped. Nobody seemed to care.

Once the food and drink are gone, my visitors parade out just as they paraded in. The children stay for much of the afternoon. We play endless games of balloon football.

They write down a number of words for me:

hattannga 'game where child follows in footsteps of somebody, hiding behind him or her';

herrvauhaq 'saxifrage';

pinngortitaq 'environment'; literally 'that that has been created'

The children gone; tranquility is regained. I sit by the window reading more MacNeice, extracts from *Autumn Journal*. All that social consciousness and poetic idiom. MacNeice is busy tracking his many selves in time. And, we too are playing that game. We are all a complex mish-mash of losses and deficits, the results of the "debris of day-to-day experience." We might like to define ourselves in terms of the division between spirit and senses. MacNeice speaks of the confusion of "this our world." I love the way he demonstrates how we always try to salvage some kind of order even when all the anxieties of that time accumulate in every conceivable direction. The 1930s. But perhaps also this time, our time.

Outside, puppies with ampersand-shaped tails race around the balcony. The dogs never go in the house. There are no pets here. The snow buntings stand in an untidy group on the tundra unable to agree on who is going with whom and for what purpose. By candlelight, I admire the palette of what might be late summer or autumn, but what is autumn without trees and russet brown falling leaves, not to mention the changing of the clocks?

September 8

Last night, the first storm of the season. It had been forecast. When the weather turns colder at the beginning of September, strong winds blow from the North (*avangnaq*). The tyranny of the wind. The *avangnaq* finds every chink in the skirting board, every cranny and cleft in the ceiling. The dining table inches across the floor. The porch door is flung open and the paraphernalia sitting on the unfinished balcony are rearranged by the wind whipping around the house. Eventually, the wind subsides and I fall asleep. I like the vulnerability; far removed from the vacuum flask of modern housing. I am seeking demonstrations of nature in the raw, and the Arctic does not disappoint. This polar desert demands that we be vulnerable. There is no 'everything happens for a reason' talk here.

The deeds of the night are behind me. I wake up to bright blue skies, high pressure and a cool breeze. Ravens patrol overhead. The blasts have carried away the snow. It is nine degrees in the house. The boundaries between indoors and out are minimal. The surrounding houses appear to be intact, and the weather looks so promising that the terror of last night seems almost a distant memory. In the bay sits a Danish frigate and in my first-thing-in-the-morning-incoherence my mind seems to think that the arrival of the Danish Navy is somehow connected to the adverse weather conditions.

I buy twenty liters (just over five gallons) of oil and am curious to see how long it will last. It cost eighty-four danish kroner or about ten pounds (thirteen dollars). Diesel is not used here as it freezes in the winter when the temperature falls below minus thirty-five degrees Celsius. I would prefer to live without all this oil. I want to find my own fuel and be properly self-sufficient. Perhaps I should have chosen the taiga instead of the tundra.

Late afternoon. I sit by the window looking at the near-hidden landscape. I wonder how it will be when it is like this all day and all night.

September 10

The sun rises in the east; a full moon sets in the west. It must be time to get up. The day starts with sunshine pouring into the hut. It is getting colder by about a degree every day. Lethargy keeps me in bed. When I get up, the temperature in the room is a chilly eight degrees. The flies have gone belly up. The strongest member of the corps dies slowly on the window sill. There is nothing I can do. It was probably as low as five or six degrees in the hut overnight. Emerging from the chrysalis of my sleeping bag in the morning has rapidly become the hardest task of each day.

I walk to the shore; the edge of the unknown. The tide is out. The icebergs feel closer than ever before. I don't tire of the grandeur of the ever-changing tableau before me. About seventy feet from the shore a grey, thin film of slushy ice is forming on the sea: the first sign of the sea freezing over. Alert to this, the snow buntings are gathering. There is a sense of urgency to their evacuation. They will soon return to warmer climes, leaving me just standing here. I will miss my conversations with them, the contented murmur of their voices. They set my mood for the rest of the day. With the exception of the raven and the ptarmigan, the Arctic winter petrifies the birdlife.

Fulmars fly just above the surface of the sea. The sun still does its fancy arc across the bay, but sits much lower in the sky now. Under the low sun, the long and formal shadows glare over the hill.

This evening, the light is quite different from previously. No longer sharp, it is achromatic. Gone are the shiny vivid colors, each with the clarity of fine enamel, of the endless Arctic summer days. Tilted-eyed dogs bay the rising gibbous, my signal to go to bed. In the not-too-distant future, they will be unshackled and will be ferrying hunters across the frozen sea.

My thoughts on today: to live alone in a hut in the Arctic, you have to be *bien dans sa peau* as the French would say. You have to be at one with yourself. Otherwise, it will be unbearable.

September 11

The diary takes a short unexpected break here. This morning, Ibbi entered the hut and told me that we were going to hunt musk oxen somewhere north of here. I had all of five minutes to pack everything for what I assumed would be a three-day journey. It turned out to be a two-week trip. In my haste, I left my diary in the cabin . . .

September 25

Back in Savissivik. The dialectics of inside and outside.

> Outside: bursting rays of sunlight between cloud shadows.
> A Tintoretto painting.
>
> Inside: already philosophizing. Trying to be an apostle of life.

I have believed for a long time that we are in danger of losing our souls. It is not just our environment that is at risk because of our reckless behavior. *We are at risk*. We are losing touch with what it means to be human because we have detached ourselves from nature, cut ourselves off from our ancient, collective psyche. Thanks to our relentless anthropocentrism, we are wandering around on a cloud of epistemological nothingness. If we want to rediscover our souls, we—the pervasiveness of humanity—will have to put nature at the center of our world again and relearn the old wisdom. We will have to learn from one another and stop pretending (in the spirit of globalization) that we have all become members of one culture. We will have to listen again to stories. And our teachers will be the indigenous people of the world. I, for one, am ready to learn.

September 28

Preparing myself for the forthcoming darkness, I am eager to maintain some of the routines I have at home. This will not be easy, but I hope such routines might add a hint of normality to what will be otherwise quite unconventional times. And so, I go for a jog. Alright, it is minus fifteen degrees, but surely there will be far more challenging hardships and adversity.

Stepping out of the relative warmth, the cold is a shock to the system. I run as hard as I can to get the blood flowing. I like to push myself occasionally. The wind stings, my face soon blanketed by sleet. Hunters stare at me as if I have lost my mind. There will be more gossip.

I run for about thirty minutes before returning to the cabin. In the evening, my ear lobes prickle and then go numb and become inflamed. The first frostbite. Lesson learnt: make sure the beanie covers the ears. I spend the evening, head cocked over the oil heater defrosting my ear lobes. I wonder what next will need defrosting.

A new word: *hiutairriho* 'has cold ears'

The author's hut pictured on the left

October 1

I am just about to climb into my sleeping bag for the night when I hear somebody coming up the steps to the house. Nobody locks their doors here, and nobody ever knocks on the door. I try to do the same. There is a bang and a tall man with glazed eyes falls into my cabin. A strong smell of alcohol. He is shouting and crashing into the furniture like a listing ship. His pleated wrinkles smile at me sardonically. A single tooth hangs down from a desperate, weathered looking face. Standing just inches away, he looks like he wants to kill me. I have no idea who he is. He must have come recently on the helicopter. The only thing I can make out from his slurring is that he wants alcohol. Firmly, I tell him I don't have any and he needs to go. He pushes past me and sits down at the table. His head soon collapses into his cradled arms; his wobbly legs twist under the chair.

He complains that it is freezing in the room and starts to kick the spluttering oil heater, shouting and groaning like a deranged man. On this occasion, the oil heater is for once my saving grace. Complaining of the cold in the hut, he finally leaves, but not before asking for two hundred kroner. He falls down the steps. A miserable pile of sorry limbs in the snow. I get him up onto his feet and quickly retreat into the cabin.

A little shaken, I lock the door, go to bed and wonder how I can avoid such encounters in the future.

October 2

The word has soon got around. Jens Ole rushes up to me in the morning and asks if everything is alright. My visitor last night is called Taiko. 'He is the only dangerous person here', he adds. 'Oh, that is a relief', I joke. He tried to kill a woman by smacking her over the head with a hammer. My mind races back to the events of last night. It was nothing really. Just a drunken encounter, I try to convince myself. But then I think what could have happened.

Without any roads connecting the settlements, criminals are flown around the country with a police escort. Taiko came on the helicopter a few days ago. He is from another part of Greenland. If you see a policeman in uniform on one of the turboprops, it is likely that he is sitting next to a criminal who awaits sentencing. Except for murderers (who are generally sent to Denmark), most criminals are sent into a form of exile as punishment. It is especially punishing for a Greenlander to be sent away from the place where he grew up. In some cases, this means ties are severed with their family, their home and their community. In Greenland, that means everything. As Greenland is such a tiny society, almost everybody will know about your crime and your name and reputation will be destroyed.

It is these small things which make Greenland unique.

In the evening, Jens Ole comes round with a gift. He unwraps a rifle and lays it by the bed. 'I hope you don't need, but just in case', he laughs. I am not sure if this is one of his practical jokes or not. 'A rifle, what do I need that for?', I guffaw. 'Well, you never know', he says. 'Keep it under the bed.' He chuckles, shakes my hand and then is gone.

Note: the convict here is so rooted in his sense of place that forced exile is a real punishment. Whereas I am doing it quite voluntarily.

October 4

I wake up to cold integrity: the oil heater went out overnight. Cold air is billowing around my head. Expletives are lost in my train of breath that lingers in the frozen hut. I cast a glance at the thermometer. Minus eleven degrees. Computer lost to the cold. I jot down some cursory thoughts in a Moleskine. The immediate question is: how much longer can I stay here? The heater is either coughing in an explosive manner and giving out minimal heat or roaring like a furnace. In February, it will be thirty degrees colder than it is now.

I go and find Qaaqutsiaq who has started to visit me on a regular basis. For a generous payment, he has offered to 'service' the faulty heater. Qaaqutsiaq, an unemployed bachelor with a constant grin, turns up ten minutes later in his overalls. *Ajor, ajor*, he says as he enters the freezing hut. 'You, real Eskimo!', he laughs. He changes the regulator and is satisfied that he has solved the problem. Terrifying grin. Before he leaves, he stokes up the fire by putting a steel rod into the heater. The rod sets fire. He runs around the hut in a frenzy, trying to put the flame out.

He is not gone long before the heater starts to choke and cough again. I spend the day taking the heater apart and cleaning the cables. It looks like I am going to be alone with this problem. Oil heater maintenance is not exactly my forte . . .

October 7

All seriousness in the cabin today. The subject-object cleavage which is the foundation for so much education in industrial societies is the source for so many of our problems. Living here, this seems to me more blatantly obvious than ever before. If nature weren't an abstraction, then we would be less inclined to abuse it. If we didn't shy away from personifying non-human living beings, then we would think twice about genetically manipulating them and subjecting them to extreme confinement. Economies of scale and cost-saving measures means that we inflict terrifying cruelty on the animals that we eat, and we feel justified in doing so because we have convinced ourselves that they do not share anything with us.

October 10

Late morning. The lethargic sun struggles above the horizon, breaking the faint Arctic blue twilight. The light is pale and wan. The low Arctic sunshine shoots a long stream of russet gold into the kitchen, highlighting small patches of crumbs and dirt on the worn and tatty surfaces. Wispy cirrus cloud hangs sublimely over the eastern tip of Herbert Island, burning out in early afternoon. The island's intriguing triangular rock formations are first pink, then orange before turning a butter yellow color.

Isblomster. Floral ice patterns. These complex, variegated October vignettes that decorate the corners of the windows sparkle in the light.

Three or four inches of snow overnight. A thin layer covers the incipient sea ice giving the impression of a flat, white landscape stretching out endlessly. In reality, the sea is a viscous, greyish soup of slushy lumps of shuga. Watching the sea ice form, this mackerel-colored façade has become a source of fascination. Every day, I peer through my binoculars charting its cycle of formation and subsequent fragmentation. Today, a solitary kittiwake flies over the thin ice with a tremendous sense of purpose.

I walk around the settlement with *Nipi* and the whip at my side. She responds well to my commands. Wooden sleds are parked outside the shop, their tall upright stanchions visible from a distance. No sign of their owners. Sleds are to Savissivik what cars are to the rest of the world. There is silence except for the crunching of snow underfoot and the distant obstreperous banter of ravens overhead. Small candles sit in the windows of simple wooden huts without lampshades or pelmets, oriels, gilded edges, gauds, lace trim or nets. There are no books on the shelf, no libretto, no Delft.

Evening falls. The darkness pours in. The land, cowered dark, holds away from me. Everything withdraws into a sepia faintness. Obscurity engulfs the settlement. Lights glitter in the cabins. It must be early afternoon. The clocks seem to have been put back. About five times.

Under candlelight, I sit late at night reading Keats's *The Eve of St Agnes*. Madeline is told that on the eve of St Agnes a virgin can, after following certain rituals, see the image of her true love:

> They told her how, upon St. Agnes' Eve,
> Young virgins might have visions of delight,
> And soft adorings from their loves receive
> Upon the honey'd middle of the night,
> If ceremonies due they did aright

Qulutanguaq, the oldest man in the community, told me just yesterday that he too saw the image of his future wife in a dream before he had even met her. As Jung would have said, we call these 'coincidences' but that is perhaps just because their causality has not been discovered. Living here in the tundra, I think I am becoming less of a rationalist! I no longer reduce anything extraordinary to the banal. I subscribe already to Einstein's "Coincidence is God's way of remaining anonymous."

The view from the author's cabin

October 13

Sunless sunsets at midday. The sky exudes soft, amber light which fades into a dark blue sky. The cliffs in the distance sit like black shag heaps against a milky, pale orange horizon.

Flanked by puppies trying to trip me up, I take endless walks. The dog teams are asleep, curled up into tight balls. Forgotten laundry sits on lines above discarded furniture gathering snow. On the beach lies a frozen harp seal (*aataaq*) like an unexploded bomb, patches of blood trace its journey up the snow-covered beach. Ibbi walks past carrying a bag of walrus meat. He killed a walrus up at Etah yesterday:

'Hello, *kammak* ('friend'), good to see you. Fresh walrus. We and the dogs are happy', he offers with a grin.

In the evening, I am summoned to Ibbi's house to dine on said walrus. Dogfights take place in the gaps of burnt sky. Lenin has broken his lead and is terrorizing neighboring teams.

Inside Ibbi's cabin. Patriarchal life. Ibbi's wife, Bodil, brings the men coffee and tea. She has been boiling the meat for two hours. In the pot adjacent to the walrus froths *qajaq*, a soup with rice, carrots and small chunks of reindeer. The walrus meat is very dark, almost black, and quite tasty. The meat is held in your left hand, and slivers are sliced off using a knife in the right hand. The blubber is also eaten, and this I do not much care for. Ibbi—a connoisseur of flippered marine mammals—talks about hunting, memory, dreams (the froth of eternity) and magical practices. Emotions flow across his face. I love his rootedness and vibrancy. He makes me feel disembodied in comparison.

Surfeited with walrus. Back in the hut, I am soon by the window again, coiled in my sleeping bag. Fascinated by the death of light. I sit until late watching the colors drain away. By the time I have finished with Dryden's *Third and Fourth Book of Lucretius*, the Roman poet who wanted us to live in pursuit of pleasure (not in fear of the gods), I find myself in the raven-cawing dark watching the battalions of shooting stars overhead. The sky is crowded tonight: satellites, planets and stars seem to compete for space. The beauty here is God's doing, not man's. With Lucretian thoughts of magnificence and pleasure washing up on the shores of my mind, I soon fall asleep:

> Which yet the nature of the thing denies;
> For love, and love alone of all our joys,
> By full possession does but fan the fire;

The more we still enjoy, the more we still desire.
Nature for meat and drink provides a space,
And, when received, they fill their certain place

A final thought for the day: now, most of us can live anywhere and believe anything. Will we at some point look back at those few who sidestepped the currents of modernity with a sense of envy?

October 15

Late morning, an unknown visitor walks into the hut. Gideon has a full head of hair, combed forward, and expressionless almond eyes. His eyes, the barometer of everything, avoid mine. His moustache is trimmed. There is stubble on his chin. He is one of the many bachelors (*qaangiunniq* 'food that has been kept too long'). He asks for a glass, removes his teeth and places the glass with the dentures by the front door. That is a first! I make him a cup of coffee. He pours all of the sugar left in the sugar pot into his cup. The whole operation is performed very slowly and with the utmost care and seriousness.

We exchange a few circumspect words. Our frozen breath lingers above the table. The conversation matter is familiar: family, dead ancestors, dreams, irregular sleep patterns etc. I discover from Gideon that the locals have set up a betting syndicate, placing bets on the date that I will leave. His coffee finished, he defecates in the infamous yellow bag and is gone. As he walks down the steps, he shouts out '1911'—he placed his money on the date of November 19. He is going to be disappointed.

Memo: there is a ridiculousness intrinsic to daily life and it can express itself in the most ordinary of gestures.

October 17

In just over a week, the sun will rise above the horizon for the last time this year. We will not see the sun again until the second week of February. There are so many unknowns, and I am not sure if I am yet prepared for this one or not. I spend more time with *Nipi* today, constantly building the bonds that unite us. Already, she understands that we are a team. She looks at me as if I were an icon, in the worship of God.

I wonder how three and a half months of darkness affects the dogs, if at all. They appear to deal effortlessly with any kind of hardship.

Titken, a West Greenlander, lost his entire dog team through cruelty and neglect. Man's vices and virtues are reflected in the animals he breeds.

October 19

Darkness and the unknown beckons. I am very curious about how the lack of light will affect me, how I will feel and how manageable this new frontier will be. Stimulation for my imagination. On 'darkness', I read Stephen Spender's *Darkness and Light* whose subject seeks through "impatient violence" to break free from the chaos of darkness and seize a more "lucid day." Slowly, I absorb the short, pithy lines. Like the rest of us, he too was perhaps groping around in the dark hoping to find his true Self and whilst at it reconcile those inner and outer worlds.

October 21

My grandmother has died. Aged eighty-six. We so often lose sight of the fact that we are a beating muscle. One day it stops, and then we are dead. In her case, she fell over, hit her head and never regained consciousness. *Memento mori.*

I sit quietly and say a few prayers. I ponder the shalts and shalt-nots. Can God still hear us if there are several hundred million people talking to him at the same time? I have no idea, but I know I am still part of this world at least. Not the 'secularist utopia', but *this* spiritual world.

October 22

Ibbi sits in my hut. The normal banter. Then 'sssh . . . .sshhh'. Ibbi can hear the walruses talking down by the shore. With these occasional utterances of his, I feel like I am an observer of another realm. What is real and what is not is a question of my own frame of mind, my own desires. Our conversation floats off into the spirit world of the mythical past and the sacred consciousness of the land.

Chapter 2: *On Darkness*

I had a dream, which was not all a dream.
The bright sun was extinguish'd, and the stars
Did wander darkling in the eternal space,
Rayless, and pathless, and the icy earth
Swung blind and blackening in the moonless air;
Morn came and went—and came, and brought no day,
And men forgot their passions in the dread
Of this their desolation

(from *Darkness* by Lord Byron)

October 24

The last sunrise. Blackout beckons. The ivory moon overhead is placed on sentry duty. The sun will not appear above the horizon again until mid-February. Yes, I know I keep saying this but it is rather extreme. A small ceremony was held to mark the occasion. The children painted pictures of the sun and a short verse was read out. The slothful sun appeared finally above the mountain at 1.32pm.

Two hours later, the sun dips below the horizon and fire spews forth. The sky glows a rich, carmine color. Dogs howl. Icebergs detonate in the distance. Countless squadrons of shooting stars light up the sky and put on quite a show. I probe them with my eyes. On such clear, dark nights the sky blinks with constellations and planets. My mind is busy photographing everything. The air is so clean; natural phenomena so immediate. In a polluted, overpopulated urban landscape, people are so far removed from these simple joys. Their cosmos is complicated, but not complex. Can they realize what they are missing? That relationship to something higher. I am so damn fortunate to be here.

Hunters are getting ready to travel on the sea ice; the winter harvest. There is a sense of anxiety for there has been such a long wait and everybody wants to get going. Down at the shore, Alutsiaq trains his new dog team, walking alongside the sled, shouting commands and expertly cracking the whip just inches above their heads. I have harnessed *Nipi* to a tiny sled and we follow Alutsiaq's example as well as we can. *Nipi* responds well to my commands, but controlling a team of twelve is a different matter entirely. A natural hauler with a pedigree stretching back over thousands of years of pulling sleds, she cannot wait to get going. I share her excitement. A small, intelligent female like her will make for a perfect lead dog, I am sure. These dogs can start pulling when they are six or seven months old and each male dog can pull up to about eighty kilograms (175lbs). The strong boys stand at the back of the pack and the nimble females at the front. A sort of inverse rugby scrum, but the hunter spends his whole time trying to avoid anything resembling a scrum.

After a few short laps along the shore ice, I feed *Nipi* some seal meat and give her a hug to celebrate her achievements. As always, she sits on my Baffin boots and pushes her head into me. We are already a team.

In the evening, I read Douglas Dunn's elegiac *The Sundial* and my mind is cast back to all those honey-colored Oxford colleges and sunken love affairs of the recent past when I was still *in statu pupillari*. All that time, love and literature beneath the gargoyles and grotesques depicting the seven deadly sins.

October 27

Early morning. Or is it? Darkness is upon us; the remaining light is being squeezed out of the sky. Another sunless sunset. Burnt clouds hang in the air like incendiary puffs. Funereal skies. The mountains in the East are as black as coal against a pale, orange sky. The wind blows the snow in sweeping arc-forming patterns. Shadows lengthen forebodingly in the moody light. The settlement seethes under the bombardment of shooting stars. It is probably noon. I am happy just watching the treasures of the world. Who else on this planet is looking up at this opera unfold in the sky?

Wistful looking dogs, the tenants of the scree slopes, watch me with quizzical expressions, heads rotating hypnotically, as I wander past in the fading light. The soft, diffused light of late October is vague and uncertain; a crepuscular incoherence. Candles flicker nervously behind frosted windows. Silent television sets, the endless flicker of this visual age, light up every room. Test Card F in cyan, magenta and blue has been playing for some hours now. The picture tears slightly, horizontally.

The slush and shuga has spawned a sea of pancake ice: a grid of colliding, frozen lily leaves with elevated rims. These plates of ice, roughly circular in shape and perhaps two or three feet in diameter creak and groan as they wrestle with one another. The boundary between sea and land is smudged by the shore ice. Icebergs of different shapes and sizes are now locked into place and sit like protruding teeth from off-white gums. A tall iceberg sits in the corner of the bay like a frozen, crumbling castle—its passage halted for many months now.

The heater is now on maximum. Twenty liters (just over five gallons) of oil lasts about three days. A low of minus thirty-one degrees Celsius is forecast for Monday evening.

October 29

I am by the window, pondering the inner truths of our age wondering whether I am a victim of civilization. I watch the shifting shadows and trembling light. It must be morning, but I have no idea of the time. The only light this morning comes from Venus (the morning star); a light bulb in the western sky. Or is it a satellite? I feel trapped between the light and the darkness: a dim cloak of different shades of grey and white that envelops the place. The sky is hoary and timeless. Drifting, swirling snow reveals terracotta colored scree. The blurred grey and white of the liminal land- and seascape merge into one like a tatty, faded postcard. The twilight and the persistent wind make everything seem hazy and indistinct: an out-of-focus sepia photograph wobbling in the water of a bathtub.

The sea ice is now forming very rapidly. In fact, there are just a few patches of open water. Minus twenty-six degrees. The collusion of the wind and the clear skies. The temperature is dropping all the time. The battle with the cold is beginning. The plyboard walls of the hut are freezing to touch. I tape up the single-glazed windows with black bin liners to prevent the cold draught getting in through the cracks. All the vents are taped up. I hang up a curtain to try and keep the heat in the living room. I now wear four base layers and an Icelandic sweater in the hut. Candles are burning everywhere; their primary purpose, a source of heat. The oil heater continues to deceive me with his unforeseen tricks, reminding me that he is Lord of the realm and not me. So, I am now anthropomorphizing oil heaters. It is going to be a long November.

In the afternoon, Qi'dduk comes to collect some walrus tusks from the loft. He will carve figurines out of them. He opens the heavy door and cold air billows into the hallway. My hut is turning into a 'cold room'. Only the contours of his face are initially visible through the clouds of freezing air filling the icy entrance. In order to get up to the loft, I go outside to collect the ladders on the balcony and my skin sticks immediately to the thermally conductive aluminum. I curse not wearing gloves and try to rip my hands away from the sticky metal. No joy. The feeling at the tips of my fingers drains away immediately. With no alternative, I walk into the hut with a set of seven-foot ladders stuck to my hand:

Ajor, ajor . . . Qi'dduk points to my hand. Spasms of laughter. Over the course of the next hour, I dangle as far as I am able my hand with aforesaid ladders stuck to it over the heater. Finally, my fingers become free and Qi'dduk gets his walrus tusks.

November 1

The sky is thistle-pink for much of the day. I sit listening to the sea ice creaking and groaning like a Shchedrin symphony singing the inhuman burden of war. The elements have been scrapping over stricken frazil crystals, but with the much lower temperatures the sea ice is thickening. There is some activity on the ice. Headtorches flashing. Dogs whining. I go to Ole's house to get a heads-up.

A dead ringed seal hangs from the ceiling in the kitchen with a pool of blood in a bowl beneath it. A young, inexpressive girl stands over the kitchen sink, bored and silent, working her way through endless piles of washing up. The television is on. We are told that there will be no weather forecast today as KNR (Greenlandic television) has not received the data from the Danish Meteorological Institute. These small things make Greenland special:

'Hunters are catching *iqalugguaq* ('Greenlandic shark') at the moment', he says.

Holes are made in the ice and they are simply hooked. These sharks are twenty feet long. They can live for over two hundred years. The flesh is poisonous, but it can be eaten if boiled in several changes of water or eaten fermented. It does not much appeal. Ole says he will go out hunting tomorrow. I ask if I can join him and he replies with an emphatic *iih*.

November 2

Last week's greyish, viscous soup has become off-white sea ice and is apparently going nowhere. It is no longer quite possible to distinguish between sea and land. The icing on the cake has set.

I turn up slightly early at Ole's house which is full of people. A *kaffimik*. The house smells of dried blood and dead animals. A small mountain of discarded boots and jackets makes entrance almost impossible. On the floor of the living room, on bits of cardboard, lie dead musk oxen and reindeer. Halibut and arctic char sit in small pots. Visitors parade in. Sat on the floor, back straight, legs at forty-five degrees like a pair of scissors, families eat chunks of raw meat, followed by cake and black coffee and then leave.

In preparation for the dog sled ride, I have donned seven base layers, a pair of orange salopettes, two pairs of Meindl socks and my Baffin boots. Ole gives me the reindeer sled skin (*inguriq*) to sit on. Clad in a sealskin parka, he unleashes two dogs which are tethered on the slope. The dogs push their heavy heads right into me. They want to play. Stood on their hind legs, they place their paws practically on my shoulders. These dogs are so strong that two or three dog teams tied together can pull a motorboat in addition to two sleds across the frozen sea. I am here to learn the tools of the trade for soon I hope I will be able to go out with my own dog team. *Nipi* will first need though some running mates.

On the ice, we collect four more dogs. In total, the team is six: five males and one female. After nearly five months of furlough, the sled dogs become a hysterical *mêlée*. The leads are attached to their harnesses which are connected to a single rope that pulls the twelve-feet sled. Instead of using a metal hook, the lines are fed into a toggle carved out of walrus tusk. The sea ice is flat and stretches out for miles before us. Whip by his side, Ole walks ahead of the dogs ordering them left and right, *harru*, *harru*, *atsuk*, *atsuk*, expertly guiding them around the thin patches of shore ice. Once we are on thicker ice, the dogs fan out and begin to run. Ole has a split second to launch himself onto the rapidly accelerating sled.

The dogs start pulling. Offensive stench of excrement. Months of fettered anguish and they are at last free to pull. It is obvious this is their vocation. The dogs attempt a semi-squat, but of course the sled does not stop for poopers. Everything is done on the move. I ask how we will navigate in the darkness and Ole produces a state-of-the-art infra-red night vision telescope.

Headtorches on, we survey the contours, lines and bumps of the frozen sea, looking for thin ice, cracks, leads (narrow fissures in the ice) etc. The only chariot on the ice. We are hunting Greenlandic shark and Ole has placed narwhal bait on large hooks at about a dozen places dotted across the sea ice. To me, the task seems more or less impossible, but somehow the dogs can smell the bait.

We travel for about four miles, navigating our way through icebergs frozen into the sea ice. I listen to the soporific sound of the sled runners bumping rhythmically across the frozen sea. It is high pressure, minus twenty-eight degrees and the sky is a pageant of constellations, galaxies and planets all trying to outdo one another.

This celestial splendor is a welcome break from the intoxicating obscurity of the Kaaba colored sky. Savissivik is little more than a cluster of yellow lights huddled together, an army of receding fireflies dwindling in the distance over the brow of a hill. Our chariot passes a city of twisted buildings made of ice: for weeks and months, distant objects through the lens of my binoculars, now up close, magnified and real.

The dogs take us from one hole to another. No sharks; the bait is still on the hooks. We return empty-handed, but I have been shown some of the networks of trails which will be my playground in the coming weeks and months:

'No problem, Stiffi. The animals and fish come to us when they are ready. They will summon us another day', he laughs, pats me on my back and hurries into his house.

On the way back to the hut, I climb the slippery slope up to the church. The light is on at the chapel. The body of a deceased must be lying there.

November 3

The darkness tells me what to do and where to go. Immersed in words, I sit in the cabin writing experimental hexameters. I tear up my scribblings, feed them to the fire, and then start over again. And again.

November 5

The wind is biting, tearing holes in my skin. There is no warmth in the earth.

* * *

The most effective way to hunt seal during the dark period (*kapirdaq*) is to lay nets at their breathing holes. That is our task today. Sat on the spider-man rug on the floor, Ole is repairing a hole in his *nannut*. After a brief conversation comprising a series of raised eyebrows and soft, drawn-out *iih*, we go straight down to the ice where we harness up nine dogs, all males except for one female. Demonic agitation amongst the dogs. There is a surge of unrest amongst the remaining shackled dog teams. Lenin and his compatriots pace up and down. They want to get to work too.

Putting a harness on a snarling, excitable male Greenlandic Dog is no straightforward business. Facing the same way, you hold the dog's head firmly between your legs and then try and work out in which hole the dog's head and legs go. Ole has never seen somebody get it right first time. I am no exception. I have brought *Nipi* along to practice the basics. She would not dream of snarling, and instead casts me a sympathetic glance as I wrestle with the straps. She sits patiently knowing that one day this should become second nature.

Glimpses of the full moon flicker through the cloud cover. The shore ice is thin and slippery. The familiar soft, scraping sound of the sled runners on the *hiku* ('sea ice'), the apparent infinities of the frozen sea and the insignificance of a solitary headtorch bumping along in the dark: the unmechanized bliss and simplicity of this kind of travel provide a nostalgic warmth to my soul. The dogs are pulling well. We travel twenty miles or so, passing haunting icebergs: ancient sculptures with artistic excrescences towering in the darkness.

Radiating from the icebergs like the veins of a leaf, leads (*ainniq*) stretch out into the white expanse. Ole jumps off and guides the dogs around the thin ice. We stop and lay three seal nets (*qassutit*). To break the ice, we use a steel rod with a pointed end. It is very hard work and we take it in turns: the ice is nearly two feet thick in this part of the bay. Almost no words are uttered. It is too cold for idle chat. Once we have made a hole, we shovel out all of the slush as quickly as possible before it starts to refreeze. We follow this procedure three times before lowering the net down the middle hole. The net is attached to two poles either side of the peripheral holes. The idea is that the seal swims into the net and gets tangled up. The seal dies because it has to come up for air every fifteen minutes. The net is not really a trap, but is more of an underwater fence pulled across under the ice. This is more fishing than hunting.

Four hours spent on the ice. Details: Icebergs the size of tower-blocks can just about be made out in the polar darkness. The open sea is apparently very close, but it seems implausible. We cannot go any further for the sea ice will soon become thin. We return in silence. Back in the settlement, the glow from the full moon lights up the whole bay. The light is soft with its soothing bluish tones. The dogs' eyes glow in the dark; their eyeshine haunting and surreal.

* * *

My thoughts for today: what does it mean at the beginning of the twenty-first century to live a 'noble' life? This question concerns me greatly. Is this escape from the complexities of modern life 'noble'? To me, here in the Arctic at least, a noble life seems to be one where man has regained the freedom of a pre-industrialized existence, a freedom that is borne from his reconnection with the natural world. More generally, a noble life is surely also one where you feel you have gained a purity in spirit, one where you have the strength to fight your cause (no matter how hopeless) to the bitter end.

November 6

A teenage boy took his life a few days ago. A bullet straight to the chin. Why? Unrequited love. The tragic plot was straight from Goethe's *The Sorrows of Young Werther*. Living in a place cut off by the winds and darkness from everywhere for months, he could see only one way out of the drama.

November 7

What do I miss from my old life? My lovers, BBC Radio 3, the chime of the chapel bell (Trinity Hall, Cambridge), the discreet clinking of plates, Sherry. I can't think of anything else at the moment.

November 9

Morning. I awake to the sound of the dogs' voices wallowing in false pathos. The oil heater is silent. Never a good start to the day. The flame has gone out and the temperature in the room is seven degrees and falling fast. The temperature is below zero in the kitchen and the bathroom. Thick clouds of breath pursue me around the hut. There is a short battle to get it going and then I go back to bed until the house has warmed up a bit. I wait for the sun to break through, but nothing. I am still adjusting to the new routine.

Today is hair washing day. I am not looking forward to the chore which is beginning to feel quite Victorian. The water in the bathroom tank is frozen. I smash up the ice with a hammer, and place the hard lumps of ice in a bowl over the oil heater. I arch my head right back, doing the limbo and dipping my head into the cold, metal washing up bowl. And, then as fast as possible, heat up another bowl of water to wash the shampoo off. This is all done in a room where the temperature is barely above zero.

A young eleven-year-old girl, Ane-Sofie, comes to visit in the afternoon. I met her one afternoon shortly after I arrived. Despite the freezing temperatures in my hut, she is cheerful and just shrugs off the cold. She does a number of drawings of me training my dogs. A European training dogs Inuit style is the stuff of caricature, and we have a good laugh at my expense. *Angir'dlarhiqtutin?* 'do you feel homesick?', she asks me repeatedly. I screw up the skin at the top of my nose. She scrapes candle wax from the candlestick holders. Boredom is a modern problem.

In the evening, I read Swinburne's *A Vision of Spring in Winter* and rejoice over the riches of the word:

> Sunrise it sees not, neither set of star,
> Large nightfall, nor imperial plenilune,
> Nor strong sweet shape of the full-breasted noon;
> But where the silver-sandalled shadows are,
> Too soft for arrows of the sun to mar,
> Moves with the mild gait of an ungrown moon

My poetry readings have become something of a sacred duty.

Re: homesickness. I have never really felt homesick. Nostalgia, yes. The further away I am from home, the more I feel nostalgic for things which we may think of as quintessentially English. The *old* England. If you are always wishing to discover the *new*, then your mind does not focus too much on the comforts of home.

November 12

The very mild weather continues. The frozen narwhal hung up in my porch begins to thaw. The sea ice is beginning to break up again with leads appearing all over the place. I take advantage of the warmer temperatures and go for a jog in the afternoon twilight.

I pass huts alive with the sound of television static, but no sign of anybody outside. The place has fallen asleep. The dogs succumbed to the allure of slumber. Old men sit by the windows smoking pipes. Content, and lost in their own little worlds. That is also a good place to be. I feel the urge to run and run. To test the limits of the penumbra, and my own strength. I head up the slope and into the couloirs of soft snow. Nothing can be heard except for the crunching under foot. I pass through cascades of stones and a landscape that has lost its imagination.

Back in the hut, I dry my clothes over the makeshift clothes line that hangs over the oil heater. There is something appealing about this simple little chore—its simplicity makes me feel grounded to something primeval. I like to watch the clothes dry over the flame. Then, I heat up two large bowls of water, strip off my clothes and have a 'body wash' using a flannel. I have to make the most of these milder days to do the jobs that are difficult or unbearable when the temperature is below freezing inside the hut. I stand naked in the kitchen. Ibbi enters without knocking. *Ajor, ajor*, he laughs. He takes a seat on the settee and waits for me to play host. Only over coffee does he tell me the reason for the visit. He and others watched me through their binoculars. *Ajorpoq, ajorpoq* ('bad'), he says, shaking his head. 'If you leave settlement, you must take rifle.' A polar bear was apparently spotted not far from Savissivik just a week ago. Ibbi puts on his ever-so-serious face and gives me a short sermon on the threat of polar bears and then takes his leave having completed his duty.

At the entrance to the hut, he turns and says: 'Stiffi. A storm is coming.' He picks up the masking tape that I left on the balcony and points it at me and then the windows of the hut. 'Be ready, *kammak* ('friend')!', he laughs and is gone.

A *kad'luna* and his masking tape versus the storms of the Arctic—the ingredients of a children's cartoon.

November 14

More terror. The wind screeches down the chimney. The house shakes. The hinged, wooden flap at the top of the steps whose purpose is to prevent the dogs coming up onto the balcony, makes a loud clacking noise increasing in tempo, faster and faster, until it is blown off its hinges and fed to the swirling scherzo. I go outside to try and get some video footage and meet Ane-Sofie. We share a grin in the squall and a smile of solidarity in the face of the elements. The wind snatches the broken words from our lips. We stand with our backs to the wind, barely being able to stand up. She loses her hat. I run after it, but it is too late and is soon skirting rooftops and chimneys. The dogs have fallen silent amidst the chaos. Further down the track, a few souls under siege, huddle and cower with coats pulled up over their heads. Like an ill-formed testudo, they shield themselves, shuffling to a nearby shelter.

The storm continues unabated throughout the day. The wrath of the skies. Thank goodness for my one book of poetry—this doorstop of philology was made for a whiteout. Qaaqutsiaq, an effigy of hardiness, has somehow navigated through the flying debris. He perches himself by the oil heater. The normal bonhomie. No need to swap pleasantries here, he launches into his normal conspiracy theory talk: the CIA blew up the twin towers, the climate change scientists' drilling into the Greenlandic ice sheet will lead it to fall into the sea and then the major cities of the world will be underwater. His views on every topic are unshakable.

A garrulous wind chats down the chimney. He, the friend of unsupported hypotheses, plays with the toy wind-up penguins (*iharukitsoq*—'that which does not have wings') that I brought for the visiting children. The toys are placed one upon the other, simulating the missionary position. Much chuckling. There is something childlike to be found in every conspirator. His constant visits comprise long, grinning silences interspersed with talk about conspiracy theories, women and sex. During such a long period of asceticism, I battle to tame my thoughts about the latter. We, men, know that the mind has more force and dignity than the body, but the temptations of the latter blur our judgement time and time again.

Late evening. I read Chaucer to dampen the libido . . . :

> On this goode wyf he leith on soore
> So myrie a fit ne hadde she nat ful yore
>
> (*The Canterbury Tales*)

November 16

I am by the window, drinking tea. I fry a musk ox steak for lunch, and spend the afternoon looking at the sky. I am magnetized by the orange tinge on the horizon. In a distant land, the drama of a sunset. Here, just the magnificent trimmings. I open the window and listen to the shore ice creaking like an old ship. The last vestige of light is disappearing from the sky. I feel a mood of melancholy press upon my shoulders. Will I lose the memory of night and day?

November 17

The orange tinge, that transient smidge of unaudited hope, has disappeared. The first day of complete darkness. The flag is lowered. How long is a day in the dark? Now that I am no longer the prisoner of time, does it even matter? Darkness is the midwife of thought; the incubator of inspection.

* * *

The main story on Qanorooq (the television news program) this evening is about an Icelandic man who has come to Greenland to sell knitted sweaters.

November 20

The temperature drops to minus eighteen degrees. The clock on my digital thermometer says it is one o'clock in the morning. I am outside with my binoculars. There is a full moon. Polar bears come closer to the settlement when there is a full moon, and I am becoming rather obsessive about seeing one. This nocturnal chiaroscuro casts a dim light over the bay, turning the sky a deep, dark blue. It is sufficiently light to be able to see stretches of open water far away in the distance. There is no wind. Without the wind, the snow sits on the ice and acts as insulation. It holds the heat in and the result is thinner ice.

Tap, tap, tap. In the afternoon, Ane-Sofie knocks on the window. She comes in and starts sobbing at my table. I have never seen her upset like this before. Not even when she loses at poker. She tells me that she is crying because she misses her friend, Sophie. Sophie has moved to Denmark. I make her some tea, and we play cards to take her mind off her troubles.

In the evening, amidst burning candles, I boil us some seal meat that I bought from Ole. Tender as butter.

November 24

I lie in the eternal darkness. No dawn, no dusk. I peep through the curtain and scour the horizon in search of an arresting wafer of light, but nothing. An upside-down world. I count the minutes, the hours, but there is no point.

My sleeping patterns are changing rapidly. Every night I am now sleeping ten or eleven hours. A lethargy is weighing down upon me. Discarding the warmth of the sleeping bag is becoming nigh impossible. 'One, two, three', I count. Then jump out of the bag, hop around the freezing floor. Compulsive swearing. I throw on as many layers as quickly as possible. With an axe, I smash up the ice in the water tank. Thaw the ice over the heater, and then cook porridge. A quick cat's lick. And then the day can begin. Same routine each day.

To take my mind off the sub-fusc heaviness, I busy myself with unnecessary errands—all rendered with a fidelity to everyday life. I spend a good hour or so cleaning the hut. I am not sure it needs it, but I feel at least as if I have achieved something.

With the twenty-four-hour darkness, I read more poetry and plunge into that private realm trying to redeem the world by introspection. Then, I am busy underlining unusual words and remarkable sentences. Heaney, Hughes, Tennyson, Muldoon, Arnold and Derek Mahon. Aloud. The constant lack of light somehow endows the words with an additional synesthetic impact. A bit like losing one sense, and the remaining senses work overtime to make up for the loss. The words of my frozen breath take on a sermonic quality as I read Derek Mahon's *Antarctica*. "I am just going outside and may be some time," indeed. Pointed understatement even in the face of death. Is that the spirit of the *real* Englishman? We should always remain calm.

A thought for my waste-paper basket of ideas: given the unhappy choice, would I prefer to die from the extreme cold or extreme heat? Which would be quicker, less painful? I spend far too long entertaining this question. I decide that dying from extreme heat would be a much longer process, and given my fascination with the North probably nothing I will ever need to worry about.

November 28

Lacunas are appearing in my diary entries. You may have noticed. I dark, my days have become unprolific. Almost ground to a halt. Today, slept for twelve hours. It was almost impossible to get out of the sleeping bag. By the time I had smashed up the ice, boiled the water, made breakfast and done the washing up, it was almost two o'clock. By four o'clock, I was yawning and my head was hurting. Ane-Sofie came to visit. Adorable girl. Like a standard flying, she warms my heart. We played cards and discussed local words such as *qammaaq* 'the clicking noise that the Inuit make with the tongue on the bottom lip to make a dog come to them'. I practically drowned myself in coffee to stay awake. Herodotus spoke of places where men slept and were awake a half-year at a time. He must have surely had the High Arctic in mind.

Today's little discovery: here in north-west Greenland, we are too far north to see the Northern Lights! My sanity was relying on these smidges of emerald.

Late evening and I read Philip Larkin's *Days*. Life might seem to bob along at a pedestrian pace, but even these days are full of self-realization and so there is a joy in the experience of living. *Days*—they wake me up, put me back to sleep and give me time to live and be happy. Perhaps Larkin who described his physique as that of a "pregnant salmon" and his face as "an egg sculpted in lard, wearing goggles" wasn't quite the sulky misanthrope I imagined.

December 1

Vaporized words linger in arid self-interrogation. There is an extended silence, a short sonata of raised eyebrows and then a meaningful grin. With quiet dignity, Ibbi sits in my freezing hut, patting his firm bulk full of seal meat. He is busy gluing multiple affixes to stems and in doing so turns long sentences into single words. Implausible palindromes are whispered over black coffee, and jokes are shared about his *embonpoint*. Thin jokes, admittedly.

Here we sit hour after hour, pensive and introspective, during the interstices of everyday life. As is so often the case in this remote, bewitching place, our thoughts turn to the urban sprawl-sopped lands beneath us, and to those especially living in choked cities. For Ibbi, living without industry ɡress' and 'development' and thus living without would prefer to live in a cosmos with a transcen- akes his head, 'we, *hamani* ['down there'], are the ⁄ou do not hunt, you do not live *in* nature', he says. hatically not one of our contemporaries. We have eality where everything is perhaps less inclined to t is reality? Reality is all possibilities.

ice. He tells me how *hamani*, there is just this mixed up, globalized w... *Inuit pa.pa.pa* . . . ('so many people, so many people'). 'There is no connection to the place, the land, the spirits that inhabit the land and now the people. You are losing your connections with one another', he continues. 'And yet at the same time there is more and more of you', he adds. An explosion in the world's population has crowded out wildlife all over the planet. We are living in denial about the biggest problem of all.

With the ice melting, the barrier to the outside world—the frozen buffer —, both metaphorical and physical, is disappearing. The skin of the world is so delicate. A time of privileged isolation is coming to an end. The world of *hamani* is coming closer and closer. Isolated from the outside world for so long, mining prospectors, Government officials, scientists and environmentalists are now entering this melting corner of the Arctic.

Ibbi does not dream of any 'elsewhere'. His *Lebenswelt* is his *nuna*, the settled land of connections, a universe of kin, family and traditions—the more obvious presuppositions. This is where he *belongs*. He has seen and fears the 'other', the *Mitsein* of nihilistic drift. When surrounded by thick crowds of *takornaqtoq* ('strangers'), he becomes the stranger himself. Tonight, the talk was serious stuff. But what connects Ibbi and I is a sense of humor, the pursuit of the light touch. We are from different worlds, but we share a Platonic lightness, even if we end up talking about *Mitsein* . . .

My thoughts in the darkness: sometimes our lives appear topsy-turvy and convoluted, torn apart from the basic modes of existence. Nature is no longer the *living* cosmos; we have forgotten its inherent meaning. We live in a fragile world. Our climate is fragile; our democracies are fragile; our friendships are fragile. We think they will last forever, but they seldom do. Oh, dear. I can see December is going to be a dark, lugubrious month.

A few more thoughts after our conversation: but this is not a 'sustainable' place either. It is part of the 'progress' world too. We are dependent on oil which comes from Denmark on a supply ship. As to its real source, who knows? We are also part of that abstract, consumerist cocoon. All the consumer goods travel thousands of miles on a polluting ship. Per capita, the carbon footprint of the Inuit is not what we might imagine it to be. Sixty years ago when they used whale blubber as fuel, it would have been zero. The problem was they wanted to live like the rest of us. And why shouldn't they?

December 3

A quiet day. I walk the length of the beach, torch in hand, looking for the words that have been orbiting in my dreams. The inner conversations that never cease. In the dark period, both time and space are turned upside down. I am present here today, but in what I do and in everything I dream of I am divided between the past and the future. I get up in the morning and part of the day has already gone, but I am not certain how much of it has gone because it is dark all the time. What remains of the day is still to come, but I am unsure exactly how much remains.

My Arctic panorama has been cruelly taken away from me, and so now I am retreating to the vicissitudes of the inner life. A streak of melancholy runs through my veins.

* * *

My thoughts this evening: so many of us choose to discard the ambiguities of the spiritual world. The fact that this seems like the most convenient option suggests we have lost the connection to the natural world, our environment. If we were integral to the natural world (as opposed to a distant observer), we would not be staring into an ontological vacuum.

December 5

Gaston Bachelard said that the "hermit is alone before God." There is a rich solitude here, and seldom loneliness. I think that rich solitude stems from a sense of alignment between nature-place-man-soul.

December 7

Morning. I stare into the looking-glass. My blond hair is ashen; my face pale; patches of psoriasis appearing on my skin. Each morning, I take Vitamin D pills. But I am beginning to crave sunlight. I am constantly tired, short-tempered and irritable. Sleep constantly overcomes me. An attack of melancholy. I am internalizing the darkness. It is leaking into my soul. The path I am on now is totally unfamiliar. What am I doing here? This kind of suffering is after all optional. Am I escaping the burden of universality? Am I hunting for an alternative to the methodological ghetto of 'progress' and 'economic growth'? Am I on the quest for some kind of impossible extreme? Or perhaps trying to fight the passing of youth? Whichever it is, I am not sure I am cut out for all this asceticism. In such a long period of asceticism, I have to constantly tame my thoughts. I am getting a foretaste of the wisdom of old age.

December 8

The cruel procession of days continues one at a time.

December 9

Thoughts about the future: reality will become blurred; language will be 'devitalized'; warfare will be cyber and digital; humans will outsource thinking to machines and digital technology; literacy will fall; the future will be techno-authoritarianism, not liberal democracy. Unless we can 'look it up', we will struggle even to think. Notes of a pessimist sitting in twenty-four-hour darkness.

December 10

George Sand said that people should be classified according to whether they wanted to live in a cottage or a palace. And what about the hermits who choose to live in a 'summer cabin' in the Arctic? They create a palace of solitude.

December 12

A storm rages throughout the day. The cabin floats somewhere between the earth and the sky. The hut trembles through me. We are united in our battle. There are few moments of reprieve. The heavens are having a tantrum. Tense negotiations with the Gods. The huts are swallowed up in a cloud of driving snow. Inside, Sibelius plays to quivering curtains and twisted walls that speak. The wind is so strong, it blows out the fire in the oil heater. This had always been the fear. Ibbi told me that in the February 2006 storm he was not able to get the heater alight for three days because the winds were so strong.

Without the heater, the column of mercury in the thermometer soon retreats to the freezing mark. Wrapped up like an onion, I read Gerard Manley Hopkins by candlelight. I always read aloud. The words vibrate in the trembling hut and unleash their secret powers:

> Into the snows she sweeps,
> Hurling the haven behind,
> The Deutschland, on Sunday; and so the sky keeps,
> For the infinite air is unkind,
> And the sea flint-flake, black-backed in the regular blow,
> Sitting Eastnortheast, in cursed quarter, the wind;
> Wiry and white-fiery and whirlwind-swivelled snow
> Spins to the widow-making unchilding unfathering deeps
>
> (*The Wreck of the Deutschland*)

Done with *The Wreck of the Deutschland*, that shipwreck and God's mercy in the midst of disaster, I listen to the demonic goings-on outside and revel in the vulnerability of the flimsy wooden hut. With my multiple layers on and lying in my sleeping bag, I am quite warm enough. But the pillow is cold, and I have to succumb to wearing a woolen hat before I can fall asleep. Despite all this, I count myself lucky to be here and am content with this kind of heroism.

I am awoken in the night by the relics of the gale.

Thought this night: coming to the Arctic tundra is a bit like visiting a cemetery. People come here to put the small and great things into perspective.

December 14

The storm has passed. I get the heater working again. The outdoor temperature has dropped to minus twenty-two degrees. Venus still sits low in the sky, above the distant glaciers. The midday sky is sprinkled with stars. I listen to the radio and pick out constellations.

On the local radio, Adolf Simigaq (yet another bachelor) says that he spotted yesterday a polar bear not far from the cemetery. I grab the torch and head off looking for bear tracks, but there is no sign of any. Still, being outside helps with the melancholy. A welcome break from the resting and the thinking. For a while at least, I no longer feel like I have lost my shadow.

Headtorch flashing, I feed *Nipi* some seal meat and walk with her around the settlement making once again sure she has learnt the commands. The darkness seems to make misanthropy more appealing. One consequence of this is that I feel closer than ever to the dogs who hold up to me the most generous of mirrors. By spending so much time with the dogs, I feel oddly that I have become more attuned to my own existence. I imagine a world with all that loyalty but without a word of verbal language to prove it or embellish it. I admire very much this loyalty of theirs, their incisive intelligence and ability to curl up and survive effortlessly the most ungodly of storms. Unlike a human being perhaps, you can pin your whole faith on a dog.

In the evening, I write poetry on scraps of paper and purple post-its. Poetic thinking has started to inhabit the cabin. The lines just come to me in these weeks of nights that leave me fondling the perimeter of my consciousness. The sounds bouncing around in my head can only be mine. All this word-knowledge. In the darkness, your imagination has to work harder to create the images. Sentences with curious juxtapositions of words just well up. The darkness has stolen that cabaret of otherworldly images—the view from my window—and replaced it with a mindscape of heaps of words, forgotten idioms and unparsed verbs. And these words are precious things. I have always believed in the sacramentality of words, the reverence of sentences.

December 16

Smashed up ice. Scraped candle wax from table.

December 17

Many of us wish to inhabit our dreams. Wasn't this Proust's revelation? But what happens if you actually live the dream? The dream dies or at least once it is here it becomes impossible to discern.

December 18

I awake to the thump of a helicopter overhead. Father Christmas has arrived. Each year, the US Air Base just over seventy miles from here and once the temporary home for eighteen thousand American soldiers, sends Christmas presents to each child. A soldier dressed in a Santa outfit climbs out of the helicopter carrying a sack of presents. For the half-dozen or so children that live here, it must be too good to be true. A small reception is held at the *katerhortarvik* ('community hall'). The name of each child is read out, and in turn they run up to Father Christmas and receive their gifts. Teary parents hug their children. Drinks and cake are served. And then Santa is on his way again to the next settlement.

This kind of interruption from the 'other world' seems quite inconceivable, but I am very happy for the children. The helicopter gone, I return to my simple daily routines, my poetry and the museum of my mind. The words are piling up in my memory. All that I cannot see in this twenty-four-hour polar night is being verbalized. I am at the mercy of words.

Reading this evening: John Donne's *Holy Sonnets*; death is just the transition between life and the after-life:

> Death, be not proud, though some have called thee
> Mighty and dreadful, for thou art not so;
> For those whom thou think'st thou dost overthrow
> Die not, poor Death, nor yet canst thou kill me

Memo before blowing out the candle and returning to the world of sleep: will digital technology render parenting a nigh impossible task? Children and teenagers will retreat to a world of inner digital experiences at the expense of external bodily experiences. Interpersonal skills will be extinguished. Young people will lose the ground under their feet. An inner reality will begin to take precedence over the external reality. The singularity of young peoples' lives will be largely internalized, and not widely known or understood by others. The period of childhood innocence will become increasingly ephemeral. The speed of the digital revolution will make the generation gaps almost unsurpassable. Older people will be exiled in time whilst young people will become exiled in place. Their sense of place will increasingly be formed through digital apps such as Google Maps, and as a result they will become attached to the world in a less meaningful and holistic way.

Despairing at what may seem to be cultural impoverishment, some will reject this digital reality and seek the sacred or perhaps the few wild places left on our planet: a transcendental experience of another kind to make their life seem more meaningful. They might even realize that in the end the secular life is not entirely fulfilling, but finding another path may be neither easy nor obvious.

December 20

I am overcome with lassitude. In the darkness I feel almost extinguished. The cabin has turned into a chamber of sleep. It is a Herculean task to get out of the sleeping bag. The hut is freezing, but the oil heater kicks into action. Literally. Tides of frozen breath.

I spend much of the day tending to my chores: washing up, cleaning the floor, collecting oil, melting ice for tea, washing clothes and drying them on the clothes line I have suspended over the heater. These chores have a cleansing effect, washing away some kind of guilt and shame. The sense of a fresh start. I revel in the glories of this poverty. It is almost a spiritual undertaking, but I am just ritualizing the innate desire to create order and harmony. Outside, in the endless dark, the world might seem higgledy-piggledy, but my little hut has to be a symbol of salvation and normality, a place of refuge and familiarity. When these lethargic days vanish into nothingness, one after the other, I have to eke out the tiniest of achievements. Like Robert Graves in his *It was all very tidy*, when he reached that place of respite of his, a lost farmhouse somewhere in the Welsh hills, he wanted the grass to be smooth, the wind to be delicate and the pictures to be straight on the wall. But then it turned out he was talking about 'death'!

Ibbi visits in the evening. He speaks of *toorngaq* ('the helping spirits that guide shamans'). Last year, he was walking along the shore when he heard a dog team come past. But nobody was driving the dogs. Then, he saw a man dressed in traditional clothing float across the ice.

Thoughts this evening: here, the pulse of life is in constant dialogue with death and the distinction between the two is rather blurred. The Inuit always have an eye on the next world. The souls of the dead are alive, skipping actively around us, reincarnated and impersonating their descendants. The transition between life and death is not the sovereign event that it is in the West. If we westerners were more spiritual, death could perhaps have a transcendental appeal. We don't all live in one universe. We live in a multiverse, parallel universes where different groups ascribe cultural meaning in different ways to different phenomena.

December 21

The shortest day of the year. And reason to celebrate. Absolutely. I can now begin the ascent down the mountain. The sky is the color of a Russian revolution. How is this possible? There has been no sun for two months, and nobody can explain this phenomenon. It is simply met with a shrug of the shoulders, and the occasional conspiracy theory. A distant atomic explosion? It has happened before.

It is very mild. The ice has melted in the bathroom water tank and there is now a pool of water beneath the ventilation flaps. It is about minus five degrees. Just a generation ago, it was at least twenty degrees colder this time of the year. It is late December, and the Arctic is almost melting. What is this lunacy? What are we to believe?

Ane-Sofie calls in on me in the afternoon. We play snap and I cook us seal meat with potatoes and onions. We toast the winter solstice and listen to the candles drip. As usual, I spend the evening fighting back yawns. The light from the waxing gibbous pokes through the small gaps of the near completely frozen over windows. I think I can hear the pack ice, shiny grey in the moonlight, creaking. Stripped of snow, the polished ice gleams in the pale glow. I have been feeling a bit out of sorts, but these tantalizing wafers of light will soon help me repair mentally. I came here for absolute freedom, and the darkness introduced severe limitations on my freedom (freedom is rearing its head as a preoccupation). I became a hostage of the night. The polar nights and the winter took away my basic pleasures: the chirping of the snow bunting, the views of frozen castles locked in the ice, the balance between night and day. But soon I will return to the winds, and the tracks on the sea ice.

December 24

I spend the morning trying to procure some reindeer meat. I promised I would cook Ane-Sofie Christmas dinner. We will be celebrating this evening. Danish style. I suggested many times that she might want to spend it with her family, but she insists on being with me. I go from hunter to hunter and eventually find nine pounds of meat. I manage to buy some potatoes and a few vegetables in the shop. I borrow some Christmas decorations from Jens Ole: The Star of David hangs in the window, some tinsel on the window sill and Advent candles everywhere.

Mid-afternoon: singing is heard. Ane-Sofie has appeared and is singing carols. She gives me a book for Christmas. It is second hand and I have no idea where she found it. She has also made me the most beautiful Christmas card. The words inside are touching and rather tragic:

'For Christmas, I just want one thing. That *you* be my father.'

She helps me finish decorate the hut, and then it is time for the Christmas service at the church. The church is packed. There is a susurrus of rising bodies as I move along the pew. Everyone is in their finery. The choir are wearing traditional dress, with *kamikker* ('fur-lined boots') and anoraks. Quite a few hymns are sung, and then there are some carols and very long readings from the Bible.

Advent red, candlelit and with windows frosted, the hut is cozy and this evening looks like an oasis in time. I love the intimacy of my refuge; the candles symbolic perhaps of a lone man holding vigil over something greater than him. My few possessions have breached the anonymity of the cabin. This little space has now been sanctified and ritualized. Dinner is not ready until rather late. The reindeer meat is very tender. After dinner: beneath a hidden moon and over the Advent crown, I read the first chapter from the gospel of St John (John 1:1-14). I open a bottle of Graham's Port, the monk that time forgot. 'Only to be opened on Christmas Day.' I remove the handwritten note. I have had the bottle concealed in the corner of one my expedition bags. It is a Ruby. The taste is crisp and intense. It practically sings on my tongue.

Thoughts on this day full of moments of grace and gentleness: I cannot help but think there is something to be gained from memorizing scripture. I have never done it, and probably never will. Memorizing is sadly no game for adults. But irrespective of whether you are a church-going atheist or a house-arrested Pentecostalist, scripture seems to represent the highest cosmic wisdom. Wisdom that has penetrated one civilization after another, and risen to the level of truth. I am struck at how terribly ignorant I am of scripture, and this ignorance appears to have become a feature of contemporary western culture.

December 25

Christmas Day. I borrow a satellite phone from Jens Ole and make a call to loved ones back at home. For a few fleeting moments, I am reconnected with that other world. In my mind, it is sometimes difficult to link the two. Actually, I have become almost schizophrenic about keeping them separate. The castaway convinces himself that he is somehow privileged for living *apart*, and he feels that this privilege is diminished, his sense of exclusion tainted, if he is in contact (acoustic or otherwise) with the place he has come from.

On Greenlandic Radio, a long list of Christmas family greetings is broadcast. In this country where so many live in remote villages dotted around the ice sheet, it is difficult for people to connect in the winter months when so many flights are canceled.

I visit all the households that have children and drop off small Christmas presents that I brought with me. We exchange greetings, drink coffee and eat too many cakes. I play tag with the children until their high-pitched screams become a little overbearing, and then I retreat to my cabin.

In the evening, I prepare reindeer stew for Ane-Sofie and I. There are a few snow flurries outside. Bach plays on the radio. The red Advent candles have been replaced with white Christmas candles: even in far-flung places and perhaps even more so in times of uncertainty or despair, we insist on the symbolic order of minutiae.

Amidst all the warmth and well-being, the hours slip by as I read us Eliot's *Journey of the Magi*. He wrote it one evening in 1927 over a bottle of Gin and shortly after converted to Anglo-Catholicism. He wanted to be part of that transition that took place nearly two thousand years ago too when the foreignness of the Magi and the Zoroastrian faith was exchanged for something that was perceived to be more compelling.

Christmas: there are so many who have long since ceased to be Christian (clergy included) but who still call themselves by this name. There are so many others who have little idea what this label means anymore, but are nonetheless the first to celebrate. I am not here to pass judgement. Who we really are springs forth from our behavior (and not our words), and on this special day I am reminded that we should act beautifully.

December 26

Some people go fox hunting on Boxing Day; others go to watch football or queue in monolithic shopping centers for discounted washing machines. I am walking around the Arctic tundra in the dark with a three-meter whip and a headtorch shouting Inuit commands to *Nipi* and *Amak* (*Nipi*'s experimental sledding companion).

After my constitutional with the two girls, I return to the cabin and get stuck into some Milton (*On the Morning of Christ's Nativity*):

> This is the month, and this the happy morn,
> Wherein the Son of Heav'n's eternal King,
> Of wedded Maid, and Virgin Mother born,
> Our great redemption from above did bring;
> For so the holy sages once did sing,
> That he our deadly forfeit should release,
> And with his Father work us a perpetual peace

All this lofty matter of reflection keeps me sane in the dark months!

December 28

As sensible as they are, the Norwegians have a word for the period between Christmas Eve and New Year. *Romjul*. A word is surely required for this fallow period characterized by the eating of Christmas dinner leftovers, a general restlessness, unnecessary dog walks and the waiting around to get New Year over and done with. None of this really applies, however, to this dark patch in the Arctic. The caribou disappeared in one evening; I am too engrossed in my poetry book to feel restless; the sled dogs drag me around the slippery settlement—my only objective is to stay on my feet and I have no idea what New Year's Eve will bring.

December 30

The temperature has dropped slightly to minus ten degrees. The weather is gusty. The relative warmth has ceded to a proprietary chill. Northern winds thrash the hut all afternoon. As per usual, the wind blowing down the chimney threatens to extinguish the faintly flickering flame. The result is that the mercury slides down the gauge; eleven degrees in the hut. Tired of wrestling with the oil heater, increasingly I am resorting to another solution: spending more time in other peoples' cabins. So, off I go to visit Ibbi.

Early afternoon. The sky is completely black. There is not even a hint of light. I just imagine that clouds are drifting across the sky.

The temperature in the hut and the darkness have become two unhealthy preoccupations. In order not to become obsessive about these two things, I need to spend my time doing the kind of activities that will give me another focus.

I have barely entered before Ibbi produces a shoulder of frozen reindeer that he hunted near Etah in the far north. We slice off slivers and dip them in soya sauce and mixed herbs. We grin like schoolboys as the meat melts in our mouths. Without doubt, I am a frozen meat convert! As always, the conversation is entertaining:

Ibbi: Stiffi, I told you about our plan for independence?

Me: Independence? No.

Ibbi: I have a big plan. *Inugguit nunaat* (northern Greenland) will become independent. We have everything we need here. We don't need the south and their politicians. The ice is melting. We have fish. We have iron and gold. The Chinese can come and mine these minerals and ship them back across the melted Arctic Ocean. This will be a new shipping lane. There are just 700 of us. We will be per capita the richest nation in the world. I will be Prime Minister, and, Stiffi, I want you to be Minister of Education. Much laughter. But he likes me to concur and not poo-pooh his extravagant ideas.

Megalomania seems to be something for hermits, I have noticed. I thought this was a place of scant ambition, but perhaps I was mistaken.

After the shoulder of frozen reindeer, we move onto some frozen halibut, eaten with strong spices and accompanied by frozen rye bread. That is a first!

I return to the cabin later in the evening. It is six degrees inside. I stare at the thermometer as it ticks down. You see the problem? It is sufficiently cold that I dispense with the ablutions, put on a few extra layers, the headtorch (my reading light) and get straight into the sleeping bag. Once I have warmed up, I thumb *The Penguin Book of English Verse* with my fingerless gloves and read aloud Oscar Wilde's *The Ballad of Reading Gaol*:

> Dear Christ! the very prison walls
> Suddenly seemed to reel,
> And the sky above my head became
> Like a casque of scorching steel;
> And, though I was a soul in pain,
> My pain I could not feel

Trains of breath chug across the icy room. Only now, it seems, am I acutely aware of the madness of all this.

De profundis.

December 31

Fireworks (*hirhaqtuq*) go off all morning long. Here I am on the tundra in the remotest place I could find, and young boys are letting off fireworks. Careful not to spill the Graham's, I put down Tennyson's *In Memoriam A.H.H.* and rush outside to see where and how this is taking place. And, then it happens.

I slip on the ice and land squarely on my back. Bang! It happened in a flash. My initial thought as I lay in pain was my trip is over. But then I hear familiar giggles and laughter. Ane-Sofie has other plans. She saw me fall, and runs over to help.

'Stiffi, Stiffi, are you alright?', she gasps.

She gets me back on my feet and I stagger back to the hut clenching my back. I am badly winded and feel bruised, but it is not a sharp pain. She makes me a cup of tea. Fortunately, the heater is working well and the cabin is cozy. I spend New Year's Eve lying flat on my back in some kind of joyful melancholy, watching the fireworks through the frosted windows. It wasn't how I had planned on spending the last day of the year, but I am happy to be part of human life and to be *trying* at least to live the dream even if there are a few hiccups along the way, even if what seem to be straight paths soon turn into labyrinths. I am here to realize myself, responding to that profound inner voice and those inner compulsions that finally refused to go away. As in Eliot's *Burnt Norton*, I wanted to explore the passages that I had not previously taken and open the doors that always seemed closed. Perhaps along the way I might find some transcendence and spirituality (that 'rose-garden' of his) and leave behind the dregs of modernity. This is all we can do, and if we become who we feel we were always meant to be, if we can emerge from the Jungian 'shadow', then death is no longer meaningless.

January 1

Here the diary takes a short break due to my back injury. The cabin is battered by one storm after another over the course of a week. The days are blank. I just accept the darkness and drink it in. I hobble around the cabin wrapped in multiple layers, read poetry, cook porridge, fry narwhal steaks, drink tea, try unsuccessfully to shave (sitting naked torso for some time in a hut where it is eight degrees is an unpleasantness that I can do without), listen to the radio, try to keep the oil heater alight, make a pair of crampons out of metal coat hangers, but do not venture outside. The occasional and very welcome visitor comes to wish me Happy New Year (ukioqtarmik pi'dduarit).

January 6

I am back on my feet. Badly bruised, but increasingly mobile. It is Twelfth Night (*mitaartut*). Candles are lit in the windows. Children are dressed in disguise as mummers, wearing black capes and grotesque masks. They go from hut to hut. There are some fireworks on the ice to mark the end of Christmas. *Nipi* and I join them. The festive chapter is closed amidst young, small friends. I am feeling very ready for a new chapter. Adventure, light, journeys, something new. Please!

My reading this evening, Auden's *For the Time Being: a Christmas Oratorio*. I think of those back at home taking down the Christmas decorations, packing up all the baubles and that annual excursion to the attic. The throwing out of the tree with its half-shed needles and the wilting mistletoe. The children returning to school and that sense of renewal. Occasionally, you have to look back and enjoy the view.

January 8

Troubled sleep. A mish-mash of dreams, but only fragments stayed with me. I dreamt that somebody was trying to break into the hut. I fall back to sleep and awake only at noon. Confused. I have slowed right down. There is nobody to judge me here, and standards have slipped as I have grown more *blasé*. I cannot remember life ever being this slow and uneventful. My muscles ache. I yawn all the time. The migraines last for days. Happiness lasts for a few minutes. As every day passes, my need for feminine society becomes keener.

January 10

I am alone in the hut. And so that private, but immense inner life starts over again.

January 12

qaamaaliaq 'month where the sun begins to come back (January) (obs.)'

There is a thin wafer of light in the sky. I begin to awake from the eternal slumber and the indeterminate shadowlands. The sun was always shining somewhere else. The temperature is minus fifteen degrees. The dogs are awake and hungry.

On the radio, there is much discussion as to why for the first time in history the sun has risen two days earlier than usual over the town of Ilulissat in southern Greenland. This has baffled everybody, but the best explanation seems to be that the melting of the ice cap has meant that the horizon is lower. If so, this would be shocking proof of global warming.

Alarming as it is, I am also excited by this news. The sun is on its way. One-by-one, each settlement travelling north will soon see the sun for the first time in months. I am beginning to count the days. I spin my own chrysalis. The dark period has been a marathon: a long hard slog where I have been motivated by one individual goal—to live this unconventional cabin life and go full circle (summer-summer). Like marathon running, I knew I would discover something about myself in the process: I take satisfaction from hardship. Toil and effort, even if a futile one, makes life meaningful. Wilfred Thesiger felt the same.

January 15

Another disturbed night. The *piteraq*, a cold katabatic wind blowing off the ice sheet, slams into the cabin. Parts of the makeshift balcony are fed to the wind. The hut trembles and shakes violently throughout the night. The thin roof starts to bulge and flap. The precious flame in the heater is blown out and the temperature plummets. The wind rasps upon the nerves. I need to go to the loo, but the thought of leaving the warmth and security of the sleeping bag is almost unbearable. In the end, bodily needs overcome mental resistance. The thick, cold air that billows from the waste bag makes it more or less impossible to focus on the job in hand.

The elements seem now to control my life entirely. There are challenges, but I warm again to the simplicity of everything here. Content with the crumbs of life. Simplicity reminds us of all the universals we live under. The virtue of cabin life on the Arctic tundra is that it purges life before filling it up.

I spend a chunk of the morning in the bowels of the heater, tinkering with the regulator. Only once I have finally got it going do I venture outside and inspect the damage. The outdoor temperature is minus twenty-two degrees but it feels significantly colder with the wind chill. The wind is stinging. The balcony is no longer fit for use; the hut's aluminum roof panels have come loose from their joints. I borrow some tools and try to fix what I can, but DIY is not really my thing. Like a trusty servant, Qaaqutsiaq comes and helps me fix the coat of armor. We make equally bad odd-job men. I cook him some seal meat to thank him for his efforts. We are soon joined by one of his cousins, and our entertainment is a noisy congeniality among comrades.

Reflecting on January and what it can bring, I read Dylan Thomas's *January 1939*:

> In the sniffed and poured snow on the tip of the tongue of the year
> That clouts the spittle like bubbles with broken rooms,
> An enamored man alone by the twigs of his eyes, two fires,
> Camped in the drug-white shower of nerves and food,
> Savours the lick of the times through a deadly wood of hair
> In a wind that plucked a goose

January 18

Everything is still again. I am out walking trying to confirm the reality of the world, the magic of the profound solitudes. These days that nobody else will remember live in me. These walks are valuable in themselves for I have chosen to do them. It is me exercising my freedom. They almost always belong to the realm of freedom and not necessity for I am never going from A to B.

I meet my new friend: the pregnant dog that belongs to one of the neighbors. Her name is *Nukka*. I gave her some narwhal meat the other day and now she wants to get into the hut the whole time. Pregnant bitches are the only adult dogs that are not chained up. When she sees me, she runs over to me and pushes her body right into mine. Like an amorous couple, we sit together for a while on the steps to my cabin. Her head snuggled under my arm.

At times, I may feel like I am alone with my mutant thought in God's antechamber but the company of dogs should not be underestimated. And I am in the thoughts of those loved ones down there in warmer climes. I receive two boxes from my darling, Rebekka. She sent me an assortment of chocolates and goodies for Christmas and then another box from Germany. I am thinking of her more than ever. When apart for so long, images of ideal and empirical reality can become confused. However, she shines forth magnificently when I open the boxes. She rounds the sky with her gleaming arc.

* * *

Thoughts this evening: a single thought can shatter us and change the course of history. The devastating wars and revolutions started as a thought that existed in the mind of an individual. A single glance from a beautiful woman can make you the happiest man alive. A single word can make you 'guilty' or 'innocent', 'rich' or 'poor', 'content' or 'deluded'. A single word can tear apart institutions. The Roman Catholic Church and the Eastern Orthodox Church parted from about 1054 over the addition of *filioque* (and the Son) to the Nicene Creed. A single word can reveal how treacherous language can be.

January 20

I spend much of the day getting things in order for I do not know what. Dishes washed; floor swept; soup boiling over the oil heater.

Nukka appears to have moved in. Not quite. These dogs cannot bear being inside. However, she has rearranged the boarding around the cabin and is sleeping somewhere under the hut. She sits there and waits for me to come back and then jumps out when she spots me. She does not stray far from her patch.

Another minus twenty-two-degree-day. I can tell how cold it is by seeing how sticky the metal front door handle is. *Nukka* and I walk to the dark shore and the interstitial shadows. This tenebrous dome will soon lift and these tired afternoons resting on stubs of time will become more normal again. This sluggishness is an intolerable curse. It is like surrendering to the inertia of the world.

Thought this evening: *iluliaq* is the word for 'iceberg' and 'great-grandchildren'. I wonder what the explanation might be? Great-grandchildren is a new phenomenon here for until relatively recently people used to die rather young.

More disturbed sleep. *Nukka* sits on the remains of the balcony howling for much of the night. Manley Hopkins in hand, I scrape away the frost on the inside of the window and look for the moonlight:

> The moon, dwindled and thinned to the fringe of a finger-nail held to the candle, Or paring of paradisaïcal fruit, lovely in waning but lustreless
>
> (*Moonrise*)

January 22

I boil *Nukka* some seal meat and place it in a bowl *in* the hut. She wolfs down the meat and then tears around the cabin sending everything flying. Despite the bitter cold, she cannot wait to get outside to her natural habitat. I am struck how these dogs so obviously belong to this environment whereas us humans always struggle to adapt.

January 24

A fire rages beneath the horizon. Smoke pours out over the Earth. This is it. There is some light. The ambassador of hope. Yesterday it was pitch black, but today the salmon-colored tinge on the horizon has returned. The afternoon arranges pastel reflections of orange and lavender on the icebergs. My improving mood is correlated perfectly with the returning light. My optimism cannot be dampened now. With this life, you become part of the season yourself. Like some kind of Pavlovian experiment, I have almost been conditioned to think that *Nukka*'s howling means the sun is on its way even if I know full well that there is at least another three weeks to wait.

The temperature has dropped to minus twenty-six degrees. At these temperatures, water forms on my eyelashes freezing them together. Like conjunctivitis. Eyelashes glued together, I go inside and let them melt over hot tea.

Now that the dark period is coming to an end, I am having trouble sleeping. My body no longer needs to regenerate. I feel like a bear emerging from hibernation.

Concerning extremes: darkness for three months is akin to living beyond the borders that limit how we know the world. At night when we sleep, we retreat to another realm. But daily life in the dark is tantamount to embracing an extreme. I am reminded of *The Unbearable Lightness of Being* where the Czech writer, Milan Kundera, wrote "a passion for extremism, whether it be politics, art or in the pursuit of a hobby, is surely a veiled longing for death." Well, I am certainly not passionate about constant darkness.

January 26

High pressure. Clear skies. The sky is turning a lighter shade of blue. My spirit is lifting. The light brings out metaphysics and fond memories. I cast a glance at the maps of my hands. The darkness has taken its toll. Patches of psoriasis now spread from the backs of my hands up my arms.

The temperature continues to drop: minus twenty-seven degrees, minus twenty-nine degrees, minus thirty degrees. The isotherms fulfil their function. To prevent frostbite, I now wear a buff every day over my nose. The hunters' bodies have adapted to the severe cold: no eyelashes, often no ear lobes.

In the evening, I throw the dirty dish water off the balcony, a drop of about eight or nine feet. The water freezes before it hits the ground. Crack! With the much colder temperatures, I am eating much more. I must be burning off so much more fat, I suppose. My body is perhaps stockpiling because it thinks winter lies ahead even if only now the days are getting longer.

The sky is the color of blood. A long, red blazing flame trails across the horizon. The first sunless sunset of the new year.

Thoughts this evening: my thinking during the dark period has been rather abstract. Living in perpetual dusk and out of sync with everything, individual traits become blurred. Details disappear, and as they do so general, broader and more abstract themes and philosophies take their place. As the contours of the world reappear, thinking shifts again from the abstract to specific details.

January 28

Nukka howls for much of the night. She is now heavily pregnant and behaving quite erratically. I have managed to get her a kennel off one of the hunters, but she wants to climb under the boarding around the hut all the time and settle there. I think I know what is coming.

Dawn reveals a proud orange tinge that flashes across the horizon for much of the day. The oil heater is bubbling away; the heat seems like a supreme luxury. I lie lapped by the fantasies of my dreams. A servant to idleness, I doze in my sleeping bag until I do not know what hour. Does time, this invention of man, matter anymore? I peep occasionally through the curtains. Hypersensitive to the slightest variations in light, I wait ever so patiently for the sun that will one day return. There is twenty minutes more light each day, and sometimes I sit in my cozy cabin counting each minute trying to internalize all the positive energy that the light will surely bring. When the heater is working efficiently, I begin again to love everything around me.

Qaaqutsiaq visits in the afternoon to share his dreams and ghost stories: voices that had laid siege to empty cabins ripped to pieces by the wind, self-opening kitchen cupboard doors and invisible hunters tapping on the window in the middle of the night. He tells me he occasionally hears a voice behind him and feels somebody tap him on the shoulder. He is struggling to get the supernatural on his side, or perhaps the darkness is just rendering us all mad. After his stories, we clink glasses. A silence reigns over the cabin. Our voices are choked. He clears his throat with an abominable sound, and then, once the darkness has soaked up the light, is gone.

I am back in my sleeping bag. More Shelley. He seemed to want to seclude himself from humanity and aim optimistically for a more beautiful *higher* world. Sometimes I sympathize with his vision but today another day crumbled away like a cliff inside of me:

> Heaven's light forever shines, Earth's shadows fly;
> Life, like a dome of many-colored glass,
> Stains the white radiance of Eternity
>
> (*Adonais*)

January 31

Minus thirty-two degrees. I am out on the ice, wrapped in Gore Tex. Just my eyes are showing; ice crystals clinging to the lids. Now, there are six dog teams stationed on the sea ice. Stooped grey forms are at work on the ice. Hunters are dressed in polar bear fur trousers, caribou fur parkas and seal skin gloves. This is not for show. Ceremonies that I cannot yet participate in are taking place: hunters are busy preparing their sleds in silence and semaphoring commands to their dogs. A civilization of skins, furs, jerry cans, Thermos flasks, frozen lumps of meat, whips and paraffin heaters.

The orange tinge is no longer a thin strip, but a broader corridor of heat that belts itself around the horizon. There is now a clear distinction between day and night. I cast a shadow and am, in these temperatures at least, alive to my fullest of being. A wintry adagio is coming to an end. I no longer rehearse the defeats of the previous day. The heart is pumping. I am excited; I feel I might even convince the sceptics that this is not just an idiotically romantic enterprise.

Whip in hand, I walk with *Nipi* and *Amak* over the hummocky shore ice, reminding them of the commands. There is a background din of howling, barking dogs but with a really firm, deep voice they walk with their heads cocked towards me, listening all the time. Soon, I will connect them up to a small sled and we will be travelling our own way over the sea ice. Just the three of us. For the time being.

February 2

A restless night, twisting and turning. High-pitched screaming sounds somewhere very close to the hut. It is difficult to know if it is a human or a dog.

I am up early, fumbling around in the dark. I locate the torch and go outside to see what is going on. The murderous cold makes my lungs contract, my face stiffen. Minus thirty-six degrees. Overnight, everything has frozen in the kitchen and bathroom: the water in the sink, the hand-washing bowl, all liquids in bottles, olive oil, toothpaste etc. Everything is frozen solid. *Nukka* soon appears from under the cabin. She is thin, starving and desperate. She has given birth, and the screaming noise is coming from her puppies.

I smash up the ice, boil some water over the oil heater and cook her as quickly as possible some seal meat. I try to get under the cabin, but the opening is too small and I am reluctant to start removing all the paneled skirting. I am not sure how long the puppies can survive in these temperatures without a proper kennel. I am told the dog belongs to the people opposite, and so I go and find Hans, a free spirit whose language describes a pantheistically animated world:

Ajor, ajor, he says. He seems frustrated by the inconvenience, but soon joins me. He wastes little time in removing the panels and smaller in circumference than me manages to climb under the hut. The high-pitched squealing and yelping pierces our ears. *Nukka* is frantic; her concern humanlike. There is a litter of five. One by one, he removes the puppies from under the hut and inspects them. He is looking at the hind legs and wants to see how they stand, whether they have the potential to be strong sled dogs or not.

He shakes his head. *ajorpoq, ajorpoq* ('bad'), he groans. He puts *Nukka* and the puppies in a kennel that has suddenly appeared by the side of his house. I wish he had done this earlier when I spoke to him about *Nukka* a couple of weeks ago. He then puts his hand in the kennel, removing two of the puppies. He holds them from their hind legs. They dangle from his fingers, suspended in fear. In a flash, he smacks the puppies' heads against the side of his cabin. Hans's face is sprayed with blood. Trails of crimson run down the wooden panels. Within seconds, they become frozen tributaries of death. He tosses the dead dogs onto the ground. Their limbs quiver for a second or two, and then hang loose. I have never witnessed such butchery. This harshness is totally unfamiliar to me:

'Better we kill the weak ones now. They won't survive in the pack', he says before stamping the snow off his boots and entering his hut.

I swallow a throatful of saliva and a little pale retreat to my cabin. I fumble around for my poetry book, hoping to find something that will help me make sense of this. I want to find perhaps a less anthropocentric attitude. Never one to gloss over the violence of nature, Ted Hughes in his *February 17th* describes the macabre details of a stillborn lamb and how a smoking body covered in soups and syrups turned out to be a yolk-yellow corpse. I appreciate Hughes's dark realism, but this morning I need just an iota of compassion, a scrap of animal-human interrelatedness. And so, I seek solace in Blake's humanitarianism:

> A robin redbreast in a cage
> Puts all Heaven in a rage.
> A dove-house fill'd with doves and pigeons
> Shudders Hell thro' all its regions.
> A dog starv'd at his master's gate
> Predicts the ruin of the State.
> A horse misus'd upon the road
> Calls to Heaven for human blood
>
> (*Auguries of Innocence*)

Thoughts this evening: birth and death is everywhere here. The injustice of this world grieves me. Where animals are concerned, it is difficult for me to be phlegmatic about such things. There is no sentimentality here. Just a very stripped-down empathy, frankness and respect. Even here in this place of peace, life is a battlefield. Was this an act of savage cruelty? Interpretations might vary. What I find perhaps most objectionable is that it should be man that wields the power, and not the puppies' mother.

Inuit philosophy: life is a struggle, and you have to get on with it yourself.

My philosophy: work hard, play hard; squeeze life like a lemon; live and let live; if in doubt, opt for Pascal's wager. I am sorry that is more a basket of philosophies!

February 4

The flimsiness of the hut and the constant sense of vulnerability take my mind back to the distant past. One or two hunters live in more modern prefabs. Curiously, when I go there my heart sinks. These huts are insulated from the drama outside, cut off from the oldest season of all. Of course, I could hardly blame them for wishing to be warm and secure, but I feel like I could almost be someplace else. When I return again to the cabin, I am somehow at one again with the world, closer to the old stories and narratives of survival. Winter in my cabin confers age upon my memories, and the relationship between the cabin and the place is absolutely real.

February 5

Outside, the sky reigns over my tiny hut. When the dark period started, I slept for hours and hours and struggled with headaches in the evening. Now, the dark period is coming to an end I cannot get to sleep at all. Worn out in the morning, but mentally alert. I spend what is left of the morning reading Robert Burns—that libidinous rascal:

> "The gust o' joy, the balm of woe, / The saul [soul] o' life, the heav'n below, / Is rapture-giving woman"
>
> (*The Guidwife of Waukhope-House*)

Back to the window. I sit here contemplating the meaning of all this for the soul. My mind tries to apply his palette of cadences and intonations to descriptions of these foreign realms, and in particular the apricot sky. Everything is now hyperreal.

In the afternoon, I go for a walk. Minus thirty-seven degrees and the coldest day so far. On my legs, I now wear two base layers and a pair of salopettes. The cold stings and penetrates the tiniest of openings. I walk to the shore. All the dogs are on the ice now. In the distance, I hear the familiar sound of a sled bumping along the ice. This will soon be *Nipi*'s and my Highway, this frozen bridge assembled entirely by nature. I will not be trapped for much longer in this tiny Arctic settlement, this incubator of my thoughts.

* * *

I love the fact that there are no machines on the ice. There is a sense of equilibrium; a sense that we (me and my dogs—loyal and afraid of nothing) are still sharing something as opposed to just trampling over it with noise and smoke. We are not here to conquer anything, but to listen, observe and respond appropriately. And I am not here to conquer the world, just the Self perhaps. I am not so much trying to return to the past either, but just wish to consecrate this present and allow it to invoke distant recollections of the past when we had to be wedded to nature for it was our bread-basket. Without nature, I lose my contours and turn to mist. I have no Promethean aspirations. I know I am not the only one to find satisfaction in these simple connections, and I take no offence in being called a 'hopeless romantic'.

February 7

There is now almost six hours of daylight each day. I am at the window with my binoculars peering through the gaps in the frost which now appear on the inside of the window. The ravens are no longer silhouettes. Outside, the air quivers in the extreme cold. I revel at the tiniest shifts in geography. I can see for miles across the frozen sea. Now, that is what I call satisfaction. In the far distance, there is the occasional tiny, wobbly speck—a hunter with his dog team.

When a fellow human being makes us feel small, we strongly object. When nature makes us feel small, we are in awe.

I wear so many layers now I resemble some kind of overweight blob. The heater is on the brink of packing up for much of the day. The temperature falls to three degrees in the hut. I am wearing all my clothes, have lit fifteen candles for heat, and am trying again to clean the oil cable that feeds the heater. There is not much more I can do now. I know little about what the course of this journey might achieve.

Thoughts this evening: images of empty places—deserts, mountainscapes, great plains—have all my life bubbled up in my consciousness as some kind of Schopenhauerian sublime. The desert's austerity provides optimal conditions for the philosopher's task. The harshness of the desert (polar or otherwise) is a purgative, a representation of some kind of spiritual journey: *Purgatio, Illuminatio, Unitio*. And, this harshness has a great appeal.

February 9

My sleep is disturbed. The pillow feels like a frozen, hard rock. Desperately cold air is attacking my earlobes. Overnight, the temperature ticks down to minus eight degrees in the hut. What is coming next?

February 11

The light is returning quickly now. The pages are no longer blank. The weather has turned milder too. I am out on the ice again. The sound of whips cracking; wooden sled runners sliding across the ice; the occasional voice. Soon, I meet Ibbi:

'Oh, my Eskimo friend. How is the igloo?', he laughs. 'You know Naimmanitsoq?' He caught polar bear on January 30. Come and visit this evening and we eat.'

On my evening visit, Ibbi—two liquid black eyes and a walrus moustache—is as jolly as always. Piles of jokes at my expense. Our faces are creased with pleasure. His good humor dispels all my gloomy thoughts. In the chaotic kitchen, the arm of a polar bear is boiling in a great big pot. The bear is served with rice. Slightly stringy in places, but the taste is very close to lamb.

Memo: there is a scruffiness to Inuit homes, but I am not passing judgement. Trying to create order in life might imply a censorious and prudent denial of our condition. We can be perfectionists, but in life we always have to live with imperfection.

February 12

The sky, that haunting twilight blue color, is etched with scribbles of light cloud. The moon has turned pale. I am off on a skiing recce. I need to test the rifle and the skiing conditions. Out on the ice, nature's solitude, that companion of spiritual exaltation, meets mine. I am happy that it is finally my turn to be a pinprick on the landscape. My turn to pass through great realms of time. From the Arctic environment I am learning modesty. The size of the icebergs and the expansive cold heaven before me make me feel humble and invite me to participate in its greatness. I don't take that invitation lightly.

The rifle strap keeps slipping down one shoulder. I make some adjustments and strap it onto my back in the manner of a biathlete. The first few hundred yards is heavy going. There is a lot of hummocky rubble where sheets of ice have collided like an earthquake-stricken road. Beyond the rubble, the sea ice is absolutely flat.

I am almost entirely alone out here. There is just the occasional dog and raven footprint. I rejoice in my freedom, and just want to absorb every detail of this polar universe. I have waited for months to be unshackled on this naked seascape, and now my mind is aswarm with floating ideas and ambitions. I want to be a custodian for all this harsh beauty.

A sled passes by in the distance. A team of dogs pulls a silent hunter across the sea ice. Polar charioteers rule the frozen sea. Now the sea ice is in place, they can play the king again. This means of travel and living is as simple as it is majestic. In total conformity with nature. On the way back, the weather deteriorates slightly and visibility is restricted. I focus on the icebergs ahead and weave my way through them, avoiding the thinner ice around the edges.

Concerning our environment and state of mind: I think if you live in a place bursting with color, say Cuba, Colombia, parts of Africa, where the sea is jade colored, flowers are in twenty shades of pink, women are dressed in violet sarongs, then you might become a painter. If you spend time in the High Arctic where the frozen sea resembles an Irish linen table cloth, starched and perfect, then you become a philosopher and devote your time to enquiries into the magnificent mystery of existence. Inevitably, you take on the colors of the world around you. You spend your time pondering basic metaphysical questions: Where am I going? What am I doing here? What is happiness? What is it that underwrites my joy? I feel rather stubbornly that I must have a clear set of answers to these questions before I leave this place. But if I don't, this won't have been all in vain and anyway I think I prefer carrying these questions through life rather than having a set of neat explanations.

A lone dog team on the sea ice

February 13

Today, the sky is pink. This means one thing: the sun will return very soon.

February 14

A very stormy night. I wake up warm. The temperature has risen twenty degrees, and a vicious wind is pumping the hut. This is the *nigeq*, the wind that blows from the east this time of the year bringing snow and warmer temperatures. The heater is bubbling away like crazy and the indoor temperature rises to twenty-five degrees. I remove all my layers and walk around the cabin in a T-shirt and pants. I never imagined this.

The *nigeq* has added a thick layer of snow to the sea ice. Welcome news for skiers and hunters. Sharp ice crystals called *pukak* can cause havoc with dogs' paws.

I go outside to get some video footage, but can barely stand up such is the force of the wind. I love being battered by the elements; the feeling of being absolutely alive in the moment. I am soon back in the hut. Make tea and watch the chaos ensue from the melting windows. In search of some prelapsarian joy, I read more Blake (*Songs of Innocence*) to escape from any thoughts of mechanistic worldviews, and jot down what I like about cabin life in the Arctic:

- Living so close to the elements
- The interest you show in nature's minutiae
- The swings between the sense of risk and vulnerability (in the context of the above) and the reassuring security and coziness of the hut
- The overwhelming relevance of the seasons and the weather to the point it becomes internalized
- The fact that noise is primarily from a natural source
- 'Ownership' is not really a meaningful word. We are not yet part of the privatized world
- The privilege of living *apart* and the perception that this separation gives you a clearer sense of things
- The lack of machines (related to above)
- The sense that I am doing what I am meant to be doing, living the life of adventure (even if it is mainly sedentary philosophizing!)

Savissivik during the *nigeq*

February 15

What does it mean to be in Greenland? The smells of blood and blubber; the sounds of the dogs barking and howling; the precariousness of life; the immediacy of nature; the sense of dysfunctionality; the feeling of 'living on the edge.'

February 16

Late morning. Naimmanitsoq, florid-faced, walks into the hut:

'Stiffi, come, we go fishing.'

Never one to turn down an opportunity to be on the ice, I throw on a few layers as quickly as possible. Salopettes go over the top. On the way down to the shore, I bump into Ane-Sofie. She wants to come and visit. I tell her that I am off fishing and suggest that she might join us. Down at the shore, Naimmanitsoq is busy getting the dogs ready. The dogs are excited. The usual chaos. I ask Naimmanitsoq whether she can come along and he says 'no.' She looks terribly disappointed. The dogs are fighting and he is losing his temper with them. We are using a full team of dogs today and it is a real challenge to tie twelve dogs together when they amount to a ball of snarling fur. Just as we are about to leave, he tells Ane-Sofie that she can come. She obviously suspected that he might change his mind which is why she hung around.

With twelve dogs and a light sled we are really moving fast. After the initial scrum, two of the dogs come unharnessed and run along by the side of the sled. They will not challenge the lead dog, and run respectfully just behind him.

We travel eastwards for about fifteen minutes. First, we navigate the chaotic jumbles of shore ice, and then we are practically skating across the top of some kind of godly cake. The sea ice leads up to the sky whose purity kills the germs of despair. The white seascape repeats itself tirelessly as if in a dream. Ane-Sofie and I sit grinning for the whole journey. Pleasure spreads over my wind-chafed face. A journey has resumed, and for that alone I am grateful. This cold paradise is just there for the taking. I love the ice. I warm to that dangerous intimacy with nature. What if the ice were to break? The ice is alive. Listening and responding to our movements across it, growing and retreating, splintering and cracking and changing color—sapphire, emerald, grey, white. Far in the distance sits an *i'duaq* ('a hut built on a sled where you can escape the cold on fishing trips'). In this part of the bay, the sea ice is about three feet thick and making a hole in the ice with a shovel is hard work. We drop a line from an industrial sized reel attached to a two-handed winch standing vertically on a metal frame in the ice. Twenty-five hooks are lowered seventy feet down into the fjord. Then, it is time to retreat to the warmth of a Primus stove and a cup of tea in a tiny plaster board hut measuring no more than seven feet square.

We leave the line down for two hours. Endless cups of tea. Sighs, groans, the soft whispering of the wind and finally Naimmanitsoq's snoring merges with the rhythm of the warbling stove. Miles and miles from anywhere, somewhere in the High Arctic. On the ice. And just white noise, that non-verbal poetry between man and the natural world. I am happy.

It is hard work pulling up the line, like hoisting a yacht's sail. We have caught seventeen halibut, some of them a terrific size and one wolf fish. The normal mayhem as we sort out the dogs' traces. One aggressive dog is mauling another. Naimmanitsoq smacks the culprit with a steel rod. The dog screams. He thrashes the whip just above their heads. They cower and huddle together, absolutely petrified:

'Hop on, Stiffi' shouts Naimmanitsoq. But the sled spins and is gone in a flash. I race after the sled. Ane-Sofie is howling with laughter. Naimmanitsoq manages to slow the dogs down to give me a chance to jump on. I should be used to it, but the acceleration of the sled always takes me by surprise. We are back in Savissivik in no time. Jens Ole sits by the shore sawing a frozen seal. The wind hisses gently.

February 17

Today, the sun should be visible above the horizon for the first time since October. Instead, it is overcast and misty. I am a little disappointed. Children gather with cardboard cut-outs of the sun, but it is all in vain. There are songs and short speeches, but as yet no sun. The boys make paper airplanes out of the song-sheets. Then, they retreat to the school, quaff all the fizzy drinks that have been prepared and engage in burping competitions. We have to hope for clearer skies tomorrow.

Chapter 3: *On Cold*

Today's a nipping day, a biting day;
In which one wants a shawl,
A veil, a cloak, and other wraps:
I cannot ope to everyone who taps,
And let the draughts come whistling thro' my hall;
Come bounding and surrounding me,
Come buffeting, astounding me,
Nipping and clipping thro' my wraps and all.
I wear my mask for warmth: who ever shows
His nose to Russian snows
To be pecked at by every wind that blows?

(from *Winter: My secret* by Christina Rossetti)

February 18

I am glued to the sky all morning. The sky is orange with horizontal wafers of iridescent pastel colored nacreous cloud. Twenty miles above us, these 'mother-of-pearl' clouds are in a league of their own. These are the clouds that inspire you to dream. The hunters say these clouds are more common in this part of the Arctic than they used to be.

At just after two o'clock, the top part of the sun, a brilliant yellow disk, is just visible above the opalescent smidges of cloud. Finally, finally, the wait is over. The sun is visible for about forty-five minutes before dipping below the horizon again. The night is now broken and the hazy, wobbly forms of the last few weeks have become the lucid borders and edges of the landscape. Light, hope and thoughts of future days. Projects are taking shape in my mind. I jog back to the cabin and pour myself a glass of Graham's to celebrate.

Completely oblivious to this ever so important event, hunters walk past the hut with shambling gait, the reality of *ennui* weighing heavy on their shoulders, and not as much as a backward glance. It is business as normal and they have seen it all before.

kiangnaai, kiangnaai 'a traditional phrase used when the sun comes back'

February 19

In the afternoon, the sky is that deep cerulean blue. And, then in the evening, quite unexpectedly the calm and silence are swept away. A snow-storm chases me into the cabin. Savissivik is shaken by a squall that lasts for about an hour.

February 20

I am in the cabin writing poetry, keeping vigil over the elements outside:

> Let me tell you about the winds,
> The *nigeq* brings a platonic white carpet of snow,
> Warmer temperatures, canceled adventures,
> Blowing in new paradigms from the East,
> Ruffling the inky feathers of the raven,
> With tinctured skies hidden behind a blurred white nothingness,
> And then the *avannaq*, a strong wind from Etah
> And the *pavanngainnaq* which comes from the North,
> Making the water flat, spilling forth,
> And perhaps the *anilatsiaq*, blowing icebergs,
> Hither and thither, thither and hither,
> And the *koororranertoq* like a *föhn* by the river,
> Mild and moderate, cool and delicate,
> And the *kanannaq* bringing mist and fog
> Until the *pikannaq*, waiting and waiting agog,
> But the hunters fear the *pigguahoq*,
> Hiding sterile promontories,
> Recalculating the profit and loss,
> Scuppering plans, silencing the Church banns

February 21

The moment the sun has returned, a couple of 'explorers' have turned up on the ice. They intend to ski up north and then across the Smith Sound to Canada and down to Baffin Island. It is very unlikely that the Smith Sound has frozen over. I hope they have done their homework and spoken first to the local people, but I doubt it. Some things don't change.

Previously, I would have had the greatest respect for these people but after having gone through three and a half months of darkness and having lived in a settlement with the local people, I feel today somewhat different. I am shocked how little interest they show in the local way of life. They just want to reach their summit and tick the box.

A few Aristotelian thoughts this evening: what are the greatest virtues? Humor, modesty, self-possession? Should we wear our erudition lightly? Our ephemeral lives should be dangerous, passionate and tender.

February 22

I am happy instead watching the sky. No longer a monotonous canvas. Clouds try to cast a shadow by standing in the path of the sun, but then dutifully stand aside. Like a behemothic dahlia, the full sun can be seen above the horizon for the first time this year. One of the great paradoxes of the Arctic is that when the sun returns, the 'cold period' starts. Light is now warmer than heat. The sun reflects off the white snow and the temperature drops. It is minus thirty-four degrees. The next few weeks will be the coldest of the year. It is eight degrees in the cabin. All I can do is light even more candles, and keep juggling the hot air between the living room and the kitchen. It is a familiar routine.

There has been no phone line or Internet access via satellite in Savissivik for six weeks now. Power cuts happen every few days. And so, I continue to remain aloof from the world, chipping away at the frost on the inside of the frozen panes inscribed with stars. I am just alone here with the dogs, writing the odd poem and watching the icebergs rearrange themselves:

> Alone, but not lonely, thinking, back and forth,
> Like a ball pushed from flipper to flipper
> In a pin-ball machine on a crumbling pier,
> Subsiding into a British post-vocation silence,
> Amidst the nostalgia of salt-sprayed slot machines

February 23

The first daylight drifts through the settlement. I am outside hoping that my few exposed pores will absorb the energy. A *soupçon* of photosynthesis will do! This feels like a win-win situation. I still feel the dark period's lethargy deep in my bones, and am increasingly dependent on stimulants of many different varieties. The stimulants between my life in the hut, and my life on the ice are now quite separate and compartmentalized:

In the cabin: coffee, poetry, candle light, boiled seal meat.

Outside: the severe cold, icebergs, rising sun, the Arctic light, the dogs' reaction when they see me, the sight of polar charioteers bumping across the ice, the sense of perceived danger from sea ice travel, vulnerability.

Sun, space and silence. After the darkness, I am obsessively interested in the tiniest changes to my view from the cabin.

I go out into the cold where my words are swallowed. I return from a walk in the afternoon on the sea ice to find *Nukka* sitting in my porch. She has broken the lead, and abandoned her puppies to be with me. She looks thin. I cook her some seal meat, give her a hug and reluctantly take her back to the kennel and fix the lead. Doing otherwise would risk her getting shot.

February 25

The night is filled with the diminishing crescendo of dogs' wailing. Good enough to fool the ears of the gods. The cold air in the hut attacks my exposed head. I feel as if my blood is barely running. Life is too tough to hand out any presents. It gets light at ten o'clock, I guess. At what is probably midday, the sky is a corn gold color. It churns out cold. The sun rises shortly thereafter and the grin begins. The splendors unfold. It now makes an arc formation instead of dropping like a penny. Once again, the sun is my time piece. I like to use the elements as a source for time; it is a more intimate link to life itself. From the vantage point of the cabin, I admire the opalescent sheen of the icebergs. These ancient lumps of ice with their varied hues seem somehow omniscient.

The sun is above the horizon for an hour or so before sinking in the west. I am by the window watching the sky spread out in apple green, excavating the dreams in my mind. Outside, beauty and harshness compete. When the sun goes down, everything dissolves like a dream. I return to the world of books. I love books. How can somebody exchange a heavily annotated book passed from bibliophile to bibliophile for a scentless, intangible e-document? The sterile life is not for me.

Blake's *The Marriage of Heaven and Hell.* Do we need conventional morality? What happens if we let it be replaced by aesthetics, as Hesse feared?:

> From these contraries spring what the religious call Good & Evil
> Good is the passive that obeys Reason. Evil is the active springing from energy.
> Good is Heaven. Evil is Hell

As poets, we like to shun dogma occasionally and think that we can stand above common morality, but that does not mean that we renounce belief.

Thoughts this evening: in the world 'down there' (*hamani*), we are hostages to the world of time. We chase it relentlessly and there is seldom any time left over. If there is, we soon become bored. Time is a trap. Here, in the Arctic cabin, there is time to watch and observe the tiniest of changes. And there is even time to write about them. In this remote corner of the world, the trivia of life doesn't crawl all over you. No need to bear the frustrations of the day for there are none. Time is aplenty, but there is never boredom for you have become part of your environment and its constant changes concern you. If you are bored, then you need to explore deeper and make sure your mind is open to new perceptions.

February 26

Qaaqutsiaq visits in the afternoon. I have not seen him since the return of the sun. *kiangnaai, kiangnaai* . . . The occasional explosion into laughter. He talks about the thickness of the sea ice:

'Only one and a half feet thick. No good. It used to be far thicker this time of the year. *Ajor.*'

Then, conversation languishes. I am wearing eight layers; he has two. And so, he is soon gone. Before he leaves, he points to the bed and asks *i'ddi quassi*? 'you, how many?' He laughs and stumbles out of the hut. [The extreme cold has blunted the libido for a while at least . . .]

The outdoor temperature is minus thirty-five degrees. These last couple of weeks I have been more or less dependent on candles alone for heat. Now,

I get through thirty candles a day at the cost of about seven pounds (ten dollars). What kind of life is this? I need to cling on for just a bit longer.

February 27

Over endless cups of coffee and bundled up in multiple Icelandic sweaters, I listen to the radio. Seven hundred miles south from here, the town of Ummannaq has finally got the sea ice but the decision has been taken to shoot two hundred dogs. If it comes at all, the sea ice is only there for about two months a year, and people cannot afford to keep their dogs for such a short hunting period. The same stories are heard up and down the coast. The ancient culture of travelling by dog sled over the sea ice will inevitably come to an end at some point.

Note this evening: these changes that are taking place are surely the canary in the coal mine. I know it sounds like a cliché and has been said many times before, but never before have such dramatic changes in the climate occurred over such a short period of time. The Arctic is melting. Our water-stricken mega-cities and the stain of industry will wash away this most ecologically sound and sustainable way of living. Those who have contributed least to the climate problem will suffer most. The siege is becoming ever more oppressive. Greenland is a shadow of things to come. I am very troubled how we fetishize economic growth, and how we couple this with notions of 'progress'. I want to change all this. We need to change not just the way we think, but the language we use to define what we are doing to the world.

Tangled paths of idealism.

March 1

I am up early witnessing the morning. I congratulate myself on living at the top of the world. I go outside to sniff the breeze and suck up all the inner, psychological enrichment. I want to know and embrace everything around me. I feel blessed by the sun; relished by the Archimedean viewpoint that I still believe this place provides. Happy with this free-range life. Sometimes I am even content with the bitter cold for it purges my more lascivious thoughts.

Snowdrifts have left the entrances to some of the derelict huts hidden by a wall of snow ten feet high. Unlike in the cities, old snow never becomes sooty heaps here. Polar bear skins flap in the breeze, hanging outside peoples' homes to dry. Eight polar bears were killed in February. There is talk of little else. The bears are found in the bays, Melville Bay in particular. The bears are never trophies; but food and clothing. Food that is shared and distributed amongst the community.

The dogs sit staring at the landscape. I don't blame them. Their contemplation seems humanlike. Seal blubber lies piled up around the place, but most of the dogs are well-fed here.

My camera will not work in these extremely cold temperatures. I want these images to last forever, but now I just have to internalize the beauty of the place and resort to memory. And so, I am out on the ice again, my eyes photographing everything. Low sun, long shadows. This time I am with the children. We play tag (*atoush*) on the ice. Tired with that, I sit them down and watch the snow on the sea ice rearrange itself. The wind draws contorted pictures in the ice, crooked smiles and bent expressions. Miniature mountain ranges, frozen castles with wonky towers are a playground for stumbling toddlers. I am not sure they quite share my interest in these things, and it is not long before it is back to *atoush*.

* * *

Mid-afternoon. An unknown visitor with a patriarchal air walks into the cabin (privacy, discretion and subtlety are notions for *outsiders*). He has a couple of rotten teeth, a sort of Hitlerian moustache and smells of blubber. This is Ingaapaluk, Jens Ole's younger brother. He came on the helicopter yesterday. His face is kind and full of the virtues of the place: light-heartedness and guileless chauvinism. Crow's feet spread to the cheekbones. His eyes, a mischievous shine. I ask him what is his business in Savissivik. Then, reflect immediately on the absurdity of *me* of all people posing such a question. He shrugs his shoulders and shows me his left hand: just two fingers. I think he is trying to tell me he no longer hunts, and is just here to be with family.

Cups of tea, coffee and then tea, he soon breaks into stories pausing occasionally to giggle. Not one for small talk, he prefers me to listen. He talks about the history of the settlement. Savissivik got its shop in 1934. Prior to that, the Inuit that lived here were semi-nomadic hunter-gatherers. The population in the 1960s was 150-200, now it is thirty-eight. He speaks of an unforgettable winter in the 1970s when the sea ice was here all year round. Provisions from the summer supply ship were ferried on sleds across the ice for a few days in July:

'We shared everything, you know. There was such a sense of community. *Iih, iih*. Oh, yes, I remember those days', he says. His throat swells with nostalgia. 'I remember 1994 too, yes, that year my father and I travelled for four days on the sea ice south of Savissivik and still no sight of the sea. Impossible now', he continues.

The afternoon fizzles away in a series of amusing and thought-provoking vignettes. *Iih, iih*, more nodding and grinning. *Taima?*, and my new friend takes his leave.

Evening. Savissivik rests in a luscious light that slowly melts away with the ease of a Schubert piano sonata. Drunk on this Apollonian landscape, I invest my time in looking out of the window and reading Geoffrey Hill. Rooted in England but with Latinate sensibilities. The denseness and musicality of his poetry. His language is almost corporeal with all those "benedictions of shadows", "mannered humilities" and "conduits of blood."

Thoughts this evening: we are all but experiential patches of films, books, photographs, friends and family sown together, an album of expressions and looks, a discography of distant, impersonated voices that plays in our minds when moments are alone, and that dances in our loaned imagination. A quilt of stitched and tailored composite identities whose whole is the sum of the parts and a little bit more. Our fundamental nature is mimetic. We speak largely in borrowed words. Life is an arabesque of chance discoveries, questionable coincidences, reflections and uncategorisable odds and ends.

When fate moves us from one place to another, this discography of voices grows louder.

A polar bear skin drying in the breeze

March 3

Hunters are shooting ptarmigans on the mountain immediately behind the settlement. This is not sport, but supper. Everybody else is playing Bingo. Whereas me, I have no idea what day it is. Heaven.

Vittus, Taateraaq's son, invites me to eat polar bear meat with him. The rhythm of life follows the cycle of nature. His cabin is chaotic and filthy, but somehow I would be disappointed if it were any other way. The coppery scent of fresh blood. Vittus—obsidian eyes, matte skin stretched taut across strong cheeks—cuts up a polar bear on the kitchen floor. His new *nannut* are blood-stained. Fighting boredom, wife and child sit on the bed which acts as the sofa playing cards. He hands me a polar bear paw to inspect.

The size of a dinner plate, heavy, padded, webbed. Twenty-four hours to boil; three weeks travelling around Melville Bay in semi-darkness with his dog team to catch.

We hold the meat and blubber in the left hand and slice off slivers using a pen-knife held in the right hand. His wife does not disturb the men whilst they are eating. Living this life is Vittus's primary concern. Stories of survival fill his mind, even if the shop will sell him most of what he needs. All of those preconceived ideas, all those antibodies that had accumulated from years of reading about life and exploration in the Arctic well up inside me:

'We need meat to survive the cold, you know, Stiffi'. He raises his eyebrows seeking agreement; and I have absolutely no reason to challenge his judgement.

The juxtaposition of modernity and tradition. Clad in bear skins, eating polar bear meat with our hands, Vittus's satellite phone rings. It is his brother, Lars. He has just caught some seal and needs some help. Vittus gets ready in a flash. I am left struggling to throw on my umpteen layers as quick as possible. With a mouth full of meat, I stumble over the great pile of fur boots in the porch and join Vittus down by the shore.

Memo: we need tradition because it gives us a sense of continuity and belonging. It reminds us that we are not accidental or arbitrary, but have grown out of something meaningful. Without this, we struggle to find our spirit and soul. And what is dignified is bound up with the inner state of the soul.

Note this evening: activist groups such as Greenpeace visit these remote parts of the Arctic. Angry that the Inuit still hunt, they come armed with a priggish morality, threats and abuse. Meanwhile, the world where they come from indulges in factory farming on an enormous scale. This is the real stain on humanity. One farm alone in Australia houses nearly 100,000 cattle kept in appalling conditions. Technology might seem like the servant of freedom, but it changes completely our relationship to nature. The Inuit relationship with animals tends to be based more on harmony and respect, and that is more than a cliché or stereotype. The moralists are headed in the wrong direction, pursuing the wrong offenders.

March 5

I love being besieged by winter. I love it when the snow climbs steadily up to the window sill. In cities, small armies of men noisily remove the snow the moment it has settled (as if it should never have been there) and in doing so alienate ourselves from what we are part of.

March 6

Awoken at seven o'clock by the storm that is blowing in. Every Sunday, the *pavanngainnaq* ('the strong wind that comes from the North') threatens to blow us off the map. A dog sled race had been planned for today, but that will not happen. *Nipi* and I have been doing a little bit of training, but hopefully not all in vain.

Mid-morning. I make porridge, chip the frost off the loo seat and then go to church. I might as well enjoy the storm. I am mainly curious to see if the three-hundred-feet walk is even possible. The gale has reduced the temperature to some kind of insanity. The wind chooses where to throw me around the settlement. Small patches of exposed skin on my face sting and tingle. The sky twists into a rictus of anguish. Curled up dogs are hidden beneath a blanket of snow. All they can do is wait. I stumble past empty huts torn asunder by unforgiving storms. Two difficulties: standing upright, and catching my breath. The wind is incorrigible; visibility close to zero. Savissivik is deserted, almost lost in a white, insidious swirl.

I stagger past the shallow graves perched on top of the permafrost. The church is empty; the silence utterly profound. I wait for ten minutes or so to see if either of the two regulars will turn up, but nothing. Should I be surprised?

A need to share the drama. And, so I go and visit Thomas who lives at the top of the settlement. The storm is demonic and unrelentless. I can just about make out his blurred A-frame house at the top. The wind is thumping into me, pushing me back down the slope. A whiteout. I lose my footing and trip over a mound of snow, near his house and just in front of his dog team. I am on the ground and snarling dogs hidden under the snow pounce at me. Now that I am on the ground, I am fair game and they take their chance. I am lucky. I am about two feet out of their reach. Otherwise, I would have been fighting off the dogs, and nobody would have seen or heard a thing.

Bare-chested and with a perpetual grin, Thomas straddles a new, full-length wooden sled parked in his living room. His laughing eyes dart around the room. The bad scar on his chin and the faulty speech. Results of an attempted suicide. He has made the *qamutik* especially for the big race that was meant to take place today. Hunters relish the competition. A question of pride and status. Thomas is one of the many *angatsuduk* ('bachelors') in the community. The women have left for the town, leaving empty houses full of memories from distant childhoods. Savissivik is a place of abandoned men (hunters and hermits) and not just abandoned houses. Occasionally, cabin fever explodes into suicide and murder. The winter before I arrived, a man had been murdered by his wife. She stabbed him with a kitchen knife. It was a week before a helicopter with a policeman aboard could land in Savissivik. The residents took orders from the police on a satellite phone, locking the murderer in the school and taking turns guarding it. The deceased (the catechist) and the wife were eventually put on a helicopter and flown out of the settlement. And, life carried on.

These men that have been left behind are the few that still want to live *in* nature, totally anchored by a sense of place. Legs and digits lost to the cold; they stay because they feel they belong here. Materialism and orderliness are their elsewhere. You can only sit on one chair at a time.

We sit, drink coffee and watch the wind. 'This can last for weeks, you know, Stiffi', says Thomas. But for most of the time we sit in silence. Speech is not the only means of conveying meaning. These long silences are a window onto a primordial, inner world with all its unspoken words and thoughts.

Question this evening: why do we write? Is it a desire to perpetuate ourselves through posterity? Montaigne was perhaps on the right track when he said:

> "Is there anyone not willing to barter health, leisure and life against reputation and glory, the most useless, vain and counterfeit coinage in circulation?"
>
> ('On Solitude', in Montaigne's *Essays* first published in 1580).

We are motivated by some kind of clandestine feeling to capture the inexpressible. I cringe when I hear that somebody has read something I have written. I feel like I want the ground to swallow me up. It was always meant to be private, and when these covert thoughts are forced into the public, I sense regret. And yet, there is a simultaneous uncanny urge to put something of yourself out there. This urge has an erotic quality, planting ideas and images into the reader's mind. For me, it seems writing never achieves its objective. Can you ever quite "catch the dewdrop and reflect the cosmos?" (as was said of Harry Martinson when he won the Nobel Prize in 1974). There are just different shades of disappointment, but perhaps secretly writers like to be misunderstood. Stalin said that writers are "architects of the human soul", but then he went on to send most of them to the gulag!

March 7

The unreality of the images outside contrasts sharply with the stark reality of life inside the cabin.

March 9

The day of the dog sled race. The storm has passed. Minus thirty-seven degrees. What wonderful spring weather! The hut is partially hidden behind Zhivagoesque snowdrifts. I emerge from the cabin like an escaped prisoner. Everything looks bright and vivid. There will be nine competitors and the distance is twenty-five miles. The course is mapped out around icebergs with a loop around the large tabular iceberg way in the distance towards Cape York. We make the final preparations and lead our eager dog teams down to the sea ice. Demonic barking and howling. All that untapped energy. The sea ice is a maelstrom of dogs. A small crowd gathers at the start. Huddled, drinking tea and jogging on the spot to stay alive. I have borrowed a few dogs. My only objective: to complete the course, and not die of hypothermia.

We are all lined up on the starting line. My *Nipi* is going to be lead dog. It can be no other way for she is the only one I have trained *ab initio*. Jens Ole has lent me a very small sled, but with all my layers I feel like rather heavy cargo. A gun is fired. The dogs get tangled up immediately, and my sled spins around in circles of incompetence. Much laughter heard from the few spectators. The other competitors accelerate effortlessly into the white distance. I sort out the tangled leads, and then head off in my own time. Afterall, there is no rush, right?

Once the initial mayhem is sorted out, things soon settle down and the dogs run well. They are enjoying themselves at least. I shout all the Inuit commands I know and occasionally thrash the whip to the left and right to ensure I stay in the tracks of the hunter ahead of me at the back of the pack. The parallel sled tracks seem to continue into infinity beyond the horizon towards some kind of cold heaven.

The cries and laughter of the spectators fade into the distance. Memories of my life in the cabin float away from me too. Back on the polar canvas, the lure of the familiar voices in my head, I commit each and every impression to memory. I know that I am meant to be here. Uncanny, raw beauty; the rhythmic shudder of the sled. Like a night train. But this time, I am sitting on the roof. The spirituality of the Arctic; it makes me reach deep into the foundations of my inner callings. Multistorey lumps of ice the shapes of chess pieces wrapped in shades of blue, green and yellow. The stuff of dreams. My main concern is not to get too close to them. *Harru, harru, atsuk, atsuk*. We weave our way through the icebergs, detritus of a distant Ice Age. My mind and body barely function: the extreme cold reduces me to the innermost marrow of life.

I keep a vigilant eye on the competition, most of whom are far ahead in the distance. This white space is their life. Our freedom on this playground is only restricted by the parameters of nature. But for the moment, I see no boundaries and if there are any physical limits, I wish to transcend them. Like the Self, the sea ice seems boundless. One by one, the hunters' calls to their dogs fall silent as I drop further and further behind. Qi'dduk, the winner, finishes the course in one hour exactly. I am a full forty-two minutes behind. By the time I get back to Savissivik, the spectators have all gone and the prize-giving ceremony is in full-flow.

March 11

My dreams were prosaic: no helicopters, dog sledding or banging polar bears on the nose with frying pans.

The *nigeq* brings the days and takes them away again. The snowdrifts climb above the window sill. The weather swallows me up. I like it this way.

March 14

Morning. The sky is empty. High pressure. Extreme cold. I can see scores of miles in every direction. Who says I am not Master of my lot? What more can you ask for? I am going to check the seal nets with Ingaapaluk—one of the neighbors in his forties. A smiley face through a haze of cigarette smoke. Close to his cabin, young dogs frolic and wrestle with one another, tossing a frozen baby seal in the air.

A happy-go-lucky bachelor. Ingaapaluk lives in one tiny room, the loo bucket is in the corner and there are two beds: one which he sleeps on, and one for visitors to perch on. It is by now a familiar arrangement. Pots and pans strewn over the floor; seal blood and intestines piled up on newspaper. Small mountains of dirty dishes and used coffee cups everywhere. There are crucifixes and faded pictures of Christ on the wall. The television is on and he is dozing. I feel embarrassed to have disturbed him, but he groans *uniit . . . uniit* ('don't worry') and waves me in. The noises in his throat are barely human. Not wishing to violate his privacy, I wait outside with the dogs.

The seven dogs (five males and two females) fight, bicker, show off and make love. I break up their little society and try and create some order. Ingaapaluk jogs alongside, whipping to the side of them and shouting orders, *ah, ah, ah, ah* ('come here, follow me'). There is the normal chaos and mayhem; high-pitched feverish excitement, growling, jostling for position and wagging of tails. I grab onto the stanchions of the sled, leaning back as far as I can to ensure the dogs do not shoot down the hummocky shore ice. When they start pulling, it is almost impossible to run with the sled. We get to the bottom of the slope without too much drama and within seconds we are on our way. It is always the same: hours, days of skulking, waiting and then suddenly, you are out there amidst all that Arctic air. A psychic release.

Hak-hak, hak-hak, Ingaapaluk navigates using the icebergs, remembering where he left his nets in relation to icebergs whose distinctive shapes make for good markers. Icebergs close to the settlements are particularly welcome because they provide fresh water, keep the fast ice in place and enable hunters to set seal nets without having to travel far. The nets are put close to the edge of the iceberg as this is where the seal comes up for air and where the ice is thinnest. As with the rest of the region, the hunters of Savissivik have noticed that the current is getting stronger. Stones are used to hold the nets in a vertical position, but hunters have found empty nets that are lying flat. The stronger current at Cape York makes it now dangerous to travel by dog sled around the peninsular, effectively cutting them off from the other settlements. In an age of mass mobility, a few of us are becoming more isolated than ever before.

We are making our way towards a *maniilaq* ('a large, high iceberg with columns') that sits prominently in front of Savissivik. I have been watching her from my cabin window. The wind has been blowing from the East for some days and as a result we are forced to travel alongside a *qimiaggoq* ('long ridge of snow formed by the wind blowing the snow in one direction'). Recently, Ingaapaluk put down three nets, dotted around the sea ice, all about three miles or so from the settlement. The ice here is about one and a half feet thick. Alas, all the nets are empty, but Ingaapaluk is his normal cheery self. *Naung-ajoq* ('what a shame'), he says repeatedly. For him, there is no time lost, no days wasted. Nature will give us food when it is ready.

Returning to Savissivik, we bear the full brunt of the *pavanngainnaq*. Minus forty-two degrees. The wind is a treacherous, guileful creature, chilling me to the bone. I keep trying to wiggle my toes and clench my fists, but I am rapidly losing all the feeling in my extremities. My buff has frozen to my beard and the tiny patches of skin exposed on my face freeze over. Sitting behind Ingaapaluk, I bury my head under my arms adopting the brace position in a futile attempt to get the blood running. A numbness strangles my body. A sharp pain anaesthetizes my brain.

Ingaapaluk wears a thinnish anorak. There is a boyish pride to him. He turns to me, grinning and says *ikkeernaqtorruaq* ('it is very cold'). Too cold to waste anymore words. How can the dogs cope? My respect for them grows by the minute. I get cramp in my left leg and feel thoroughly disabled and disembodied by the time we get back to Savissivik. A lesson in humility. When Chekhov was in Siberia, he wrote of how the wind lashed his face to "fish scales." I know the feeling. I limp off the sled as we approach the shore ice. The dogs race up the hummocky ice, the sled speeding off towards his cabin. Even for a very experienced hunter, they are extremely hard to control at this point.

Guffaws of laughter. Ingaapaluk sees me hobbling. What is the news? Qi'dduk has killed a polar bear at Cape York. An excitement presides over the place. I thank Ingaapaluk and retreat to my cabin to regain my body and senses.

A pale light in the evening. I am back on my feet, marveling at the sky. The moon is reflected in the windows of the abandoned yellow cabin behind Thomas's house. Both derelict and charming. The light has a fugitive beauty. The cabin has been ripped apart by Arctic storms; its shabby state is inexplicably photogenic. The sinews of silence. No suburbs here. That sense of timelessness again where it is almost as if the present has stopped going by into the past. No idea which day it is.

Then, the crunch of the hard snow as I stroll past half-dozing dogs who have one eye on me. Their snouts are slightly raised, identifying me by smell. Broken furniture and discarded white goods peek up through the thick snow: the accoutrements of modern day living scattered across the frozen tundra.

* * *

Notes on killing of polar bear:

First *tikeraat* ('a bear that comes into a settlement')
in Savissivik since 2006

Male

Ten feet long

Six-digit number on gum; tracked by biologists

Thin, almost no blubber. Highly dangerous

Shot with two bullets

Meat shared

No trophies

Three men take twenty minutes to skin it

Not known how many bears in the Kane Basin

Older hunters agree polar bears used to be much fatter

I retire at just gone midnight, but am awoken at 2.50am by Qaordloqtoq, one of the younger men who helps out at the store. He is standing at the entrance to the living room, shining a torch in my face, and it is obvious that he is excited about something. He tells me that I have to come immediately as there is a polar bear in front of my cabin. Confused. Is this some kind of practical joke? It would hardly be the first one. I curse at how slow I am at getting ready.

We run outside into the pitch black. Minus thirty-five degrees. I have my pocket video camera, but no torch, and stumble with sleepiness in the deep snow. In my confusion, I forget for a moment that I am in a hunting community and have in my mind this image of people watching a polar bear walking majestically past. Sadly, I am mistaken and the truth is remorsefully other. A dead male polar bear lies directly beside my cabin. It was shot two minutes ago. He had smelt the seal blubber lying around and had taken his chance. The hysterical barking of the dogs woke up my neighbor and the animal was killed quickly. Ever since I arrived, I have been going on and on about wishing to see a polar bear, and thus Nukappiannguaq's first words were to wake up the *tuluk* (the 'Englishman'). A small crowd has gathered around the bear.

The adult male bear (*ittorruk*) is nearly ten feet long. One swipe of its massive paws could kill a human being. Ropes are tied around the bear's legs and it takes five men to drag the bear six feet up the slope where an electric lamp has been strung to a pole. The bear's jaws are opened. A number of its teeth are missing on one side and its jaw is broken. This is mating time and it looks like he has been fighting with another male.

I watch the goings-on with a sense of shock and despair. This is the first polar bear I have ever seen in the wild. Here it is: the dead mass before me conjures up images of large prehistoric animals. The bear is cut up. I am used to all the blood and guts by now. Nukappiannguaq is excited like a boy and very proud of his kill. I don't share his excitement. The bear is turned on its back and a small cut is made just below its chin, then the knife goes right down its stomach. A bear is cut up in the same way a seal is. The men are freezing, trying to cut up the bear wearing just thin rubber gloves. Once the bear has been skinned, tea and coffee are served inside Nukappiannguaq's house.

If there were ever any doubts about the kind of community I am living in, then now there are none. Not sure what to make of the whole experience, I go back to bed.

March 17

Forty Licks floats out from my cabin. I am reading Hardy. Not an obvious combination. Thomas Hardy's (*The Voice*) makes me think of a relationship slipping away, of all that uncertainty piled up in a sweet disorder. The distance between us and the long silences is beginning to hurt:

> Can it be your that I hear? Let me view you, then,
> Standing as when I drew near to the town
> Where you would wait for me: yes, as I knew you then,
> Even to the original air-blue gown!

The moon is full; a lunar eclipse. Its silver rays flood the settlement. I have never seen such a serene sky. Such an "imperturbable serenity." It is bitterly cold but I am happy to go outside to freeze and enjoy the splendor of the syzygy. Far from the Hardyesque continents of moil and misery, this profile is indeed as "placid as a brow divine".

March 18

I wake up to sunshine and blue skies. A ptarmigan (*aqiggiq*) stands facing the wind, pecking at the ground apparently oblivious to both me and the extremely low temperatures. The wind blows the snow off the slopes, throwing up small puffs of white clouds. The view over the sea ice resembles the Sahara el Beyda, the white desert of Farafra: the icebergs protruding like white desert mushrooms created by the dry wind and the sand; a magical, frozen Kingdom whose ice is wrinkled like parchment.

Trying to photograph the bird, my fingers freeze, go numb and rapidly start to sting and burn. I am unable to put the lens cap on and decide to dive into Adolf's hut to avoid imminent frostbite. He is busy making a sled. He puts his chisel down and welcomes me in. I dash over to the heater and slowly thaw my frozen fingers over the flame. He quickly puts some coffee on. The routine is by now extremely familiar. Adolf (Thomas's brother) tells me that they eat ptarmigan here and that it tastes very good. 'Similar to chicken, Stiffi'. Like his father, Taateraaq, he is a warm, kind soul living the simplest life possible with almost no material possessions. Ripped bin liners cover up the windows to prevent the severe cold seeping in through the cracks in the woodwork. He does not care for envy, outdoing his neighbors or for socially beneficial cycles of competition.

Adolf, an active hunter, has spent his whole life out in nature and has at times been victim to the hostile, cruel climate. He made the costly error of falling asleep on his sled—the ultimate nightmare for so many hunters. He woke up far from Savissivik in the silence of the pre-dawn morning. His dog team had stopped and were lying down. His left leg had frozen; a lump of ruptured cells that barely still belonged to him. I think I have some sense of the terrible fear and agony that he must have gone through. Today, he has a wooden leg and life goes on as usual for him. A stoic, he greets each day with a grin and is out hunting whenever he can. These hunters are the real Stoics. Just three generations ago, it is said that elderly people who believed they had become a burden for their families because they could not hunt would roll off the back of sleds and die of exposure. There were no teary goodbyes.

March 20

It feels like the first day of spring, but I know that it can't be. Seasonality here is not the normal four-course affair. It is a full ten degrees warmer in the sunshine and for the first time in weeks it is possible to have a conversation with somebody outside for more than a few minutes. It is a novel sensation after so many months, not to be freezing. I help Jens Ole build an igloo for one of his dogs to give birth in. The snow is compacted and very thick. Large blocks of snow are cut with a saw and make for bricks. The whole operation is remarkably simple. It is pleasing to see so soon the fruits of our labor.

Another bear has been killed. Children run around the settlement, knocking on each door and spread the news. To celebrate, a *kaffimik* is held.

In the evening, I do some dentistry. I set about repairing a molar which broke off last week and has been causing some discomfort. Using a tiny handheld mirror, I 're-cement' the tooth using my emergency dental repair kit. No idea whether it will hold.

A reflection: this tiny grain on the map might seem to epitomize insignificance. But what happens here is of great relevance to the rest of the world. This huddle of cabins on the edge of the ice sheet is not just an annex for my dreams. When the sea ice melts completely, the settlement will probably die. The fishing is poor here. Nobody has boats. It is costly to insure them. And so, the dispossessed hunters will move to the town to join the other brigades of broken men. The subtle ways the landscape shapes the Inuit will be different; the journeys that led inwards to the Self will have to be reconfigured; dogs will be shot in their hundreds; hunters will be displaced and become stay-at-homes; traditions and stories will be lost to the wind.

March 21

Assaulted by another storm. The *nigeq* terrorizes Savissivik day and night. Four thousand years ago, meteorites fell here. Now, we are caught in some kind of polar vortex. I sit in the cabin eating frozen arctic char and reading Pope. As early as 1728, he was writing about England's cultural decline.

March 23

Late morning. Qaaqutsiaq stops by. More conspiracy theories and intriguing falsehoods: we should respect the Taliban for their conviction; temperature drops when the sun comes back because the sun is so far away. It gets warmer because the sun moves closer to Earth; man-made climate change is a myth; cremation prevents reincarnation.

March 25

I love the light. Clarity of mind, as well as vision. There is now about fifteen hours of daylight each day, and I am so happy to study the language of the sky. The occasional day is surprisingly mild. In the afternoon, the sun spills into the hut pushing the temperature up to a torrid twenty-five degrees. It reminds me of how it was when I first came here. The weather is changing fast. I remove all the bin liners from the windows and open the vents. It feels like an historic event.

I am outside without a hat. First time in months. I am used to covering obsessively every square inch of exposed skin. The change feels sudden, almost too sudden, jarring perhaps.

A young man saunters past, stripped down to his T-shirt. The temperature is minus fifteen degrees.

March 27

The brilliant Arctic sunshine spews over the clear skies above us. The iceberg in front of my cabin window shines like a precious stone. The shape of the sea ice is changing all the time. My thoughts and consciousness struggle to keep up with it. Overnight, the ice rubble fields have more or less disappeared, reconfiguring the visual poetry of shade, light and line:

> The ice was here, the ice was there,
> The ice was all around:
> It cracked and growled, and roared and howled,
> Like noises in a swound!
>
> (Coleridge, *The Rime of the Ancient Mariner*)

I put down the poetry and go to meet the light. I am soon on the ice amidst all the paradisical beauty. Ice fishing with Ibbi. At least half of the population are out here flicking their wrists, hoping to entice a polar cod or two. Jokes and biscuits are passed around. Shovels are at the ready. Keeping the holes free of ice is a constant battle in these temperatures. Ibbi talks about the men that are hunting walrus, over by Northumberland Island at the edge of the ice sheet. A seriousness invades his face. I can tell he admires the walrus hunters' grit and courage. Notions of heroism come to mind. Walrus hunting is the most dangerous form of hunting in this corner of the Arctic. When he talks, he has one eye on the sky watching how the wind scatters the cloud to see if the *avangnaq* ('a northerly wind') is headed in our direction. The tundra has seeped into him.

One hour, two hours. I am overwhelmed by the boredom of it all. Thank goodness for the sun halos overhead. I have caught one polar cod. A small fish, measuring about eight inches long. It will be supper.

Two facts regarding death in Greenland: (1) if you want to have your body flown home to your settlement or town, you have to have a zinc coffin in case the body fluids leak; (2) there is not enough room in a helicopter for a coffin, so the coffins are strapped underneath.

April 1

Today I read Charles Wesley, skied for three hours, caught a polar cod, watched a seal through the binoculars, made tea and porridge, sucked on narwhal blubber, discussed the winds with everybody I met, planned a trip to Northumberland Island, listened to the *Rolling Stones*, collected oil and finally fed the dogs.

April 2

More Wesley, less narwhal blubber, fewer seal.

April 4

> Sweet Day, so cool, so calm, so bright / The Bridal of the earth & sky! / I see with joy thy chearing light, / And lift my heart to things on high
>
> (Charles Wesley, *Hymn for April 8, 1750*)

Ice. Ice everywhere. I walk on the sea ice. Mouth open, I taste the wind. Salty, grainy. Memories of childhood. I swallow the occasional snowflake. Garnish for the soul.

I drill a hole in the ice with my hand-held auger. Tired of polar cod, now I am looking for arctic char. Two hours later. Arctic char [0]; catfish [1]. The 'earth-quake' fish. This mythical, bottom-feeding oddity tastes as bad it looks. I spend the evening searching my mouth for bones.

April 6

More planning: maps, compasses, first aid kits, satellite phone call trials, conversations with hunters.

April 7

hila—'the local cosmos, mind, consciousness, climate'

I awake to a sallow light. Sky painted with a few strands of wispy, feathery cirrus cloud in the east. The hardship of the winter is behind me. Hemmed in by solitude and terrorized by four months of darkness in this unforgiving land, I intend now to make the most of my freedom. So, I have been busy codifying my intuitions.

Sunshine cascades into my weather-beaten hut twenty-four hours a day. The season of moonlit ice is over and the regime of time will be of little relevance for the next few months at least. It will only start to get dark again in late August. There is no reason to travel to Northumberland Island (*Kiatak* in the local language), this uninhabited island in north-west Greenland to where I am headed. It is just there, staring at me in the far distance on those cloud-free days when you can see for a hundred miles. Most of the time it is hidden in a haze of speculation. Remote, and frequently visited by polar bears, it is a place where walruses forage from sea ice platforms and arctic hares pair off to their mating territories. For me, the fact that it is an island, a place on a map is good enough reason to go there. An adventure that obeys the coordinates of some kind of predestined journey. Something more than just *vagabondage*.

Unwrapped, the possibilities of this frozen world seem endless and no matter where I am in the world, I always have to turn over every stone. I am drawn inexorably to the ice. The appeal of this silent world is spiritual; my internal landscape merges with the unimpeachable authority of the external landscape in an all-embracing vision. I have been searching for something greater than myself, a corner of the world that has not yet been rationalized. What it yields to the eye is more than a panoramic view, it is an extension of the human psyche. There is beauty in a landscape which solicits mental freedom. There is a clarity of perspective gained from standing at the top of the world. The sea ice invites a kind of solitude that can be filled with the consciousness and grandeur of these spaces. Up here, the thought of loneliness seldom enters my mind. Instead, I harness the land and sea for sustenance. The satisfaction of self-reliance. Simple daily chores are my essential liturgies. I am living slowly again.

The Arctic brings everything down to the core: when you are nothing but a tiny, insignificant pin on the frozen sea, the contours of life stare back at you from your long shadow cast in the low April sun. You need to survive and you will only do so if you correctly interpret what surrounds you, what you have become part of (*hila*).

Northumberland Island aside, I construe the edge of the sea ice as my true destiny. I imagine this frozen corridor, the Murchison Sound, tapering to infinity, continuing beyond the icebergs that wobble in the haze in the unquantifiable distance. I feel like a flat-earther with my wish that the ice would continue *ad infinitum*. I want to meet the sun. And so, I had to undertake this journey, to travel to the edge, to see if there was an edge at all and to see what was there on the boundaries of the human psyche.

The sled has been packed and repacked rather obsessively: fuel canisters, oil heater, shovel, frozen halibut, walrus and seal meat, *tooq* ('a wooden pole with a pointed edge, a bit like a harpoon but used for testing the thickness of the ice'), *inguriq* ('a reindeer skin to sleep on'), rifle, *pituqqamavit* ('spikes screwed into the ice when the dogs are tethered'), binoculars, satellite phone with solar powered batteries, Thermos, kettle, tea and chocolate etc. We will soon be ready for off. The sled is ready, waiting with the dogs on the broken-up shore ice. I leave my hut perched at the top of the slope and walk down to my mini dog team. Sleeping dogs stir as I walk past, trying to assess the chances that I might have food for them. Down by the shore, a small crowd is gathered to see me off. Their faces are all framed in my mind now as if that moment is paused and photographed in my memory. They know better than anybody how capricious and fatal the Arctic can be.

I keep the dogs tethered individually and use a triangular fan formation with *Nanoq*, a lead dog (*ittuqut*) that I borrowed from Jens Ole for the sled race. The two bitches (*Nukka* and *Nipi*) will not undermine his leadership. *Nanoq*, humble, attentive and nine years old now, can still pull an 80kg (175lbs) weight. With a tremendous amount of fur on his head, *Nanoq* (the name means 'polar bear') is very bear like. The leads are attached to their harnesses and through the walrus tusk toggle are connected to a single rope which pulls the twelve-foot sled.

As I unleash the dogs, jealousy boils over and the dogs begin to fight. Scolding the dogs for their rivalry and love interests, I shout orders (*aulaitsit*) to the scrum of snarling fur caught in the spaghetti of leads. I swish the whip an inch or so above their heads, the dogs cower and soon the social hierarchy is reinstated. We need to move fast now before the fighting begins again. I walk in front of the dogs, holding the whip, and shouting *hak-hak* ('get going'). Once we get onto the flat sea ice beyond the shore, the dogs suddenly accelerate and I have a split second to launch myself onto the sled. For this split second, there are three options: time your leap right and launch yourself onto the sled, let go and watch your dog team disappear into the distance or worst of all miss your window, hold onto the sled and be dragged along the sea ice behind it. Thankfully, this time it went to plan and I give those on the shore line scrutinizing my preparations little reason to gossip. We all know that I am an amateur at this game.

The dogs are set free. I too feel their excitement, that familiar alloy of fear and elation at being let loose in this frigid house of sky. The dogs soon get thirsty. They deftly drop their tongue to one side of their jaws and eat the snow as they run along the ice. There is the occasional extremely rapid semi-squat, a response to the excitement. The particles from the excrement's steam freeze instantly forming chandeliers of crystals that become wedged in my beard. The conditions are good. A slight wind from the north corrugates the surface snow on the sea ice. It is minus twenty-seven degrees. We weave our way through the icebergs for hours.

Our silhouette soon advances westwards along the coast. I stay the night at the tiny hunters' hut at Cape York. I had spent enough time staring at its location on the map that it proved not so difficult to find. A pile of 1980s pornographic magazines in the corner. The hut is empty. The dogs are tired, and I set about feeding them as soon as possible. I chew on snacks of frozen halibut, put on my snowshoes and with rifle over shoulder, set off to photograph the sculptures of ice.

Thoughts this evening: the journey that I have undertaken today will probably be impossible in ten to twenty years' time. The sea ice will have melted. Further in the future still, the sea ice of the Arctic Ocean will melt. Then, the Chinese will come looking for iron which they will ship directly across the open water of the Arctic Ocean. The Northern Sea Route running along the top of Russia could become a major shipping lane. The Arctic could be industrialized and militarized. Antarctica is protected by the Antarctic Treaty which forbids all these activities. My wish is that there were such an international treaty to protect the Arctic. I suspect it is wishful thinking, but it is painful to contemplate whether what I am experiencing now will be gone forever.

April 8

The fine weather continues. I am up early with the sun, and wish to make the most of the excellent conditions. The day greets and embraces me.

The sky is first pink, and then the soft orange light of the early morning sun. The view is intoxicating. The mountain ridges have a surreal sharpness as the folds form an alliance with the light. Ahead of us lie statues of ice leaning like drunks on lampposts made of antique glacial water. The dogs are running well and we push on at a steady pace keeping clear of the thin ice around these icebergs. I read the wind from the *aijupinak* ('striae or thin grooves') on the ice. The snow has been largely smoothed over here by the *nigeq*. I scan the sea ice for polar bear tracks (*tumi*) or evidence of any other kind of life. There are long trails of raven footprints etched into the ice like a secret hieroglyphic code. Expertly camouflaged ptarmigans explode from the snow.

We reach a trail of enormous pinnacled icebergs parked in the slow current zone running westwards; their transit through the Murchison Sound halted until the summer months. Their edges and grooves shine and glisten in the morning sunshine. These frozen guardians stare down at the solitary dog sled and grin at our toil. The silence is very nearly absolute. It seems to reveal to the soul that which the spirit desires. The stillness of the eternal beginning. The world as it had always been.

I measure our breaks by the icebergs and decide that we will stop when we reach the smallest, flat-topped tabular iceberg (*natsinnarraaq*) resembling a monumental frozen coffee table. It turns out this particular iceberg is colossal. It just appears small because it is so far away. Hours later and we have still not passed it. Some of the icebergs we are passing are like mountain ranges or ancient, turreted castles. It is almost impossible to guess the height of the icebergs as there is nothing to compare them with, but some might be over three hundred feet high. In the Arctic, there are icebergs nearly the height of the Eiffel Tower.

The light is constantly playing with my mind, flashing me tantalizing distorted images of weather-beaten prefabs floating in the sky: the optical pitfalls of the Arctic. Northumberland island is suddenly without fixed definition. My eyelashes glued together with *kaniq* ('the frost that forms on eye lashes'), I am not sure I can trust my own eyes.

ai, ai, ai, ai, I shout to the dogs. We stop for a rest. I listen to the crooning wind. I thaw my eyelashes over a mug of steaming tea. The dogs lie in the sun, their protruded noses sniff the wind. I watch the *natserivik* ('the snow that blows across the surface of the sea ice') create, reshape and then erase mesmerizing patterns on the ice. Within a few weeks, this frozen highway will suddenly transform: the ice around the western tip of Herbert Island will become thin because of the *aukarneq*, the fast current that runs around the peninsula; leads will begin to appear in the ice indicating the movement of walruses underneath, glaucous gulls will congregate on the sea ice. The island's slopes will come alive with the arrival of coalitions of squabbling seabirds.

Beyond the largest icebergs, the smooth sheet of sea ice turns subsequently into a major ice rubble field (*manii'dat*) reflecting the strong current underneath; a chaotic frozen jumble of broken ice stretching for miles. This basically *is* Caspar David Friedrich's *Das Eismeer*.

Instead of practically skating along, the dogs now scramble over sheets of collided, fractured ice jutting up about five feet high. I jump off and push the sled from behind, leaning with all my weight against the stanchions. The sled groans and creaks as it is thrown at obtuse angles. The next hour or so is spent fighting our way through a polar obstacle course. We manage not to flip the sled and to get beyond the rubble field and back onto the flat ice.

Our journey continues for nearly ten hours under the halo of the brilliant sun. The sound of the sled runners bumping rhythmically across the frozen sea is soporific. Exhausted from pushing the heavy sled over the rubble field, my head begins to dip. Fortunately, my legs are a bit too long for the sled. They keep sliding off the wooden platform, jabbing into the hard sea ice under the moving sled, reminding me that I should not doze.

Out of the sun, the temperature plummets suddenly ten degrees and it becomes desperately cold. The moisture from my breath freezes on my face. My eyelashes begin to freeze together again impairing my vision. Practically anaesthetized on the sled, I force myself to hop off every ten minutes or so and jog behind, holding onto the uprights and trying to keep the blood circulating to my feet which now feel like lumps of wood.

By evening, we are facing Northumberland Island. It feels as if we are travelling to a borderless mythical place, the end of the world, in search of the elusive ice edge. It is late by the time we arrive at the hut on Northumberland Island. The dogs clamber up a steep slope and there at the foot of a mountain is a tiny, gingerbread colored hut which Nukapiannguaq built in the 1980s. The roof is meringued with snow. There is a network of huts around this part of the Arctic. Many have been damaged or blown away in storms, but thanks to four massive boulders attached to the roof with ropes, this one is still standing.

In preparing for the trip, Nukapiannguaq invited me to use his *qingnivik* ('a subterranean meat store'). This is a food cache located under the ice where hunters store meat. When hunting polar bear, hunters are away for weeks at a time and cannot carry all the meat that they need. The *qingnivik* is an outdoor freezer; a hole covered with a large slab of ice which polar bears are unable to move. As the temperature is below zero for much of the year, meat can be stored here for many months at a time. Using my *tooq* as a lever, I prize open the lid to the *qingnivik* and grab a frozen leg of walrus meat.

I shovel away the snow that has drifted against the entrance of the cabin and open the door with the polar bear proof handle. I curse at the rubbish left behind by the previous occupants, but quickly clean everything up. Then, I suspend the frozen lump of walrus meat over the brass Primus paraffin pressure stove using a pulley system conveniently installed in the hut. The meat slowly thaws. I feed the dogs. Armed with binoculars and a rifle, I go for a short walk to see if there are signs of polar bears. Tonight, for the first time during my stay in the High Arctic, this feels like a real possibility. It is nearly midnight. Wisps of snow skitter in the distance. The sky is empty. I am no longer among men. I stand, enveloped in inviolable silence. There are different types of silences, but this silence is an unforgettable one: a harmonious feeling of being alone, but being party to something bigger than myself. The meaning of life is almost somehow revealed. I am content to have finally set foot on, what is for me at least, my Arctic Nowhere. Content that I have been given the strength to go my own way. I have withdrawn from this world, and for a short while bask in the silence and solitude as an abstract spectator.

* * *

I stroll back to the cabin. All is intensely still. The crunching sound my boots make as I walk on virgin snow echoes around the place. Everything seems to echo profoundly in the cold, both past and present. Even if it seems I am alone with the dogs, as I walk up the slope I can see there is evidence of other life here. Delicate, oval shaped tracks of the arctic fox crisscross the snow. By the time I get back to the hut, the dogs are curled up, sleeping in tight balls of tan, copper and sepia. My bed for the night is a reindeer skin on a hard, uncomfortable wooden sleeping platform (*i'ddeq*). I fall asleep to the sound of dripping walrus meat. It is nearly midnight and the sunshine engulfs the cabin.

Memo: Alexander von Humboldt understood these acoustic perceptions when he described the increase in volume in lower temperatures. In a lecture to the Academy of Sciences in Paris in March 1820, he said:

> "The noise is three times as loud by night as by day, and gives an inexpressible charm to these solitary scenes. What can be the cause of this increased intensity of sound in a desert where nothing seems to interrupt the silence of nature? The velocity of the propagation of sound, far from augmenting, decreases with the lowering of the temperature. The intensity diminishes in air agitated by a wind, which is contrary to the direction of the sound; it diminishes also by dilatation of the air, and is weaker in the higher

than in the lower regions of the atmosphere where the number of particles of air in motion is greater in the same radius" (reproduced from Alexander von Humboldt and Aimé Bonpland's *Personal Narrative of Travels to the Equinoctial Regions of America* Vol II: 264, published in 1852)

April 9

I awake in this fortress of constant, sempiternal light. It feels like I am sitting under a spotlight. The brilliance is dizzying. I am more accustomed to the dim light of candles. The heater is still warbling and the hut remained warm overnight. I peel back the reindeer skins, prepare some porridge for breakfast and check on the dogs. In absolute silence, they sit on the lookout. Everything is glossy and precise. With a lighter sled, we will be travelling further west today towards the edge of the sea ice. Having spent some time repacking the sled, we travel for about two hours before coming close to the end of Northumberland Island. Hakluyt Island (*Apparhuit*) is visible in the distance, but is surrounded by open water.

Just beyond the Kissel glacier, I stop the dogs (*ai ai ai . . .*) to test the ice with the *tooq*. There are grey patches and the pole soon goes through the ice. Clouds of smoke (*pujoq*), the shape of ogees, above the water in the distance tell me that we are near open water (*imaq*). The sea steam is like a laundry room. This so-called 'sea smoke' occurs when still cold air overruns the warm, moist air at the sea surface. Leaving the dogs on the ice, I head to land and climb up to a vantage point, binoculars around neck and rifle slung over shoulder. Ahead, the ice is thin with patches of open water between major ice rubble fields. The recent full moon must have resulted in a lunar spring tide which has broken up the ice. The sea ice that forms during neap tides is normally firm and smooth. That is when the Inuit hunt walrus. There will be no neat, clean distinction between ice and open water, as I imagined. I will not be tiptoeing up to the edge of this white world and looking out onto a pale blue sea. There are no edges, clear boundaries or black and white certainty about what lies beyond at all.

With the binoculars, I scan the smashed-up ice. It squeaks and creaks like an old rusty door, indicating that it is thin and dangerous. The dogs can smell something. Their nostrils are flared and they are looking west. In the distance, I can hear a deep grunting noise. I look in that direction with the binoculars, but see nothing. Then I spot a walrus lying on an ice floe next to some freshly broken ice. It is a male, and the grunting noise is its mating call. I sit down in the snow and through the binoculars watch this giant mound of blubber and hide wobble on the ice. He awkwardly drags his bulk across the ice floe, just to flop into the water and then climb slowly back up again. Close to where I left the dog team, I spot some *nakkut*, narrow leads in the ice which have frozen over again, and possible indications of the path of the walrus moving underneath towards land. There must be other walruses around, but I cannot see any. Satisfied that we can go no further, I rejoin the dog team and we travel along the edge of the rubble field to see if there is any other sign of life before heading back to the hut.

Once at the cabin, I climb up the slope to the look-out-point. I spot a sled way in the distance: an almost indistinguishable smudge on the silvery mantle. Iggiannguaq, a hunter who lives in Savissivik, turns up at the hut two hours later wearing polar bear trousers and a caribou skin anorak (*qulittaq*). He has a kind, gentle face. His wrinkles smile at me, forming a network of narrow channels and tributaries meandering between low banks of skin across his forehead. Shortly after he arrives, he digs up a large slab of meat from the deep subterranean refrigerator. In a hail of crude jokes, he soon joins me in the hut. Over a pot of boiled walrus and blubber, Iggiannguaq tells me about his views on life *hamani* 'down there'. Never having set foot in a town, let alone a city, he shakes his head and says *ajorpoq* ('bad'), 'too many selfish people violating the harmony of existence'. He talks to me about how the doctrines of *hila* determine the course of the day, the consciousness and mind and how nature's delicate pleasures make you content, fulfilling the poverty of your inner life. I don't need convincing. These last weeks I too have been touched by that spiritual aristocracy and humility.

We talk, pour tea and watch through the window the wind reupholster the sea ice. Anxious to escape the stifling heat of the hut, I cut up the thawed seal meat and go and feed the dogs. Smelling the blood and guts slopping around in the buckets, the dogs bark and howl; the normal feeding-time commotion ensues. Their cries and songs escalate into a crescendo that circles around the island. It is only after feeding time and during the worst weather that the dogs are completely silent. Later in the evening, their baying resembles a form of evening communion to which I fall asleep.

The dogs: an eclectic fellowship of saints and sinners whose personalities shape every event.

April 10

I awake to lavender northern skies. I repack the sled and make the preparations to head back to Savissivik. Shortly before I am about to head off, Nukappiangguaq calls me on the satellite phone. He wants me to bring back some meat from the *qingnivik* for his dog team. I load up the sled with slabs of frozen walrus. I say goodbye to Iggiannguaq who sits in the cabin, sharpening his knives. He hugs me and says *nanngmanniartutin aggurruaq* ('make sure you look after yourself'). There are tears in his eyes.

On the way back, I see polar bear tracks that lead from Herbert Island northwards across the Murchison Sound. The tracks are fresh and clearly defined, but no sign of a bear. The journey to Cape York seems never-ending at times and the dogs are tired pulling our swollen load. The weather has been kind to us again. I know how very fortunate this is. I cross myself and thank a higher authority. We all sleep soundly.

Thought: in this place where there are no trees and the light is so clear, space, time and energy have meanings that are different from those in a delineated, evenly modulated temperate zone. I am living in another dimension. Outside of time. Time and space are not cut up into small, regulated parcels, but are open and fluid, merging with one another.

April 11

Late evening. Back in Savissivik. It is as if we never left. A group of hunters are on the shore ice, sharing jokes whilst repairing a sled. As I approach the settlement, the men start to clap. One of them bellows: *piniartoq nuutaq* ('the new hunter'). Everybody laughs. I talk with the hunters about Northumberland Island and then feed the dogs. I produce the frozen walrus meat from Nukappiangguaq's *qingnivik* and tell them what a successful hunting trip I have had. *Hunaa* . . . ('oh, gosh . . . '), one of them shouts and then they roll around laughing. I walk up the slope to my hut. The cabin is cool, but not freezing. The oil heater belches as it always does when the tank is low. I refill the tank just in time and the room starts slowly to warm up. I am exhausted and get straight into my sleeping bag. I peep through the frustratingly thin curtains. The sun circles above me in a sky without contrails. Then, the satellite phone bleeps. Immediately, I recognize the voice that speaks to me in slow, creaky Polar Eskimo:

'Stiffi, it is Iggianuguaq. The broken ice has been all blown away. I am at the ice edge now. There is just the deep blue sea and I can hear the walruses talking. You should come back, my friend. There is time. We wait for you.'

Chapter 4: *On Light*

Busy old fool, unruly Sun,
Why dost thou thus,
Through windows and through curtains call on us?
Must to thy motions lovers' seasons run?
Saucy pedantic wretch, go chide
Late schoolboys and sour prentices,
Go tell court huntsmen that the King will ride,
Call country ants to harvest offices;
Love, all alike, no season knows, nor clime,
Nor hours, days, months, which are the rags of time

(from *The Sun Rising* by John Donne)

April 12

Sleeping has always been my forte and today I sleep late. Very late. I make porridge. Life is a feast that begins with breakfast. Ane-Sofie soon drops by and wants to hear about my trip. We drink tea whilst she scrapes candle wax from the table. A diligent scraper. A very familiar routine. The operation takes the best part of an hour. She grins at my travel account, permanently perplexed that a *kad'luna* is running a dog team and living off frozen fish.

Noise: a distant rumbling becomes louder. We play cards and chew on halibut. This roaring, thumping sound overhead. The heavens shake. We rush outside, puppies at our heels. The cold stings and sharpens my spirits. A Hercules flies right over my little cabin. The return of machines. Echoes of a previous life. I feel like it wants to trample on our insignificance. Here we go again. A weakness for allegory. The plane—a turboprop cargo monster—seems larger than the settlement itself. It is on its way to the Thule

US Air Base—a Cold War construct built for another ideology. A reminder that even here we are somehow part of this modern world.

Fears regarding cabin life:

- Perceived remoteness is 'undermined' if symbols of 'other world' are introduced
- Noise from machines
- Exiles of a similar disposition turning up
- The perceived 'threat' represented by the over-populated 'other world'
- Misplacing book(s)

April 14

There is such strangeness hidden in the ordinary. Just take language: all those silent letters of the alphabet; suppletive verbs (go-went); the Icelandic verb *gleyma* ('to forget') which used to mean 'to be merry'; irregular plurals (ox-oxen); that strange German word for 'destiny or fate' *Schicksal* (literally, 'sending to the room'); all those verbs in Russian with two forms depending on whether the action is complete or not etc.

April 15

I am on my way to Herbert Island. I want to see the hut where the explorer, Wally Herbert (the island is not named after him. Just a strange coincidence), lived. He crossed the Arctic Ocean in 1968 and lived on the island to learn the art of running dog teams. Then, there was a small community on the island but the settlement closed down in the 1990s when they stopped delivering oil.

This time, I am on skis. Roped together like mountaineers, I leave the dogs in Savissivik. I feel an urge to push myself physically, to use up all the energy humming through my body.

The first smiles of a pink-colored sun. Alone out on the Arctic sea ice. Never happier. I have full faith in the white carpet beneath me. I am navigating myself through an antediluvian snarl of captivating shapes of ice. All have been molded by the wind and the melting process, making their shape transient and ephemeral. I know this journey will never be quite the same again. Next year, the sea ice will be different, the icebergs unexpected shapes and sizes. Other countries have monuments, castles, trees and cathedrals. Here, we have ice trying to adopt the shape of all these objects.

Mid-morning. Herbert Island's mountain ridges have an unusual sharpness as the folds catch and collaborate with the light. After some hours of huffing and puffing, I see in the far distance what appears to be the abandoned settlement where Wally lived in the 1970s. Now, a jumble of memories swirling in the wind.

The rucksack is horribly heavy. I curse at having brought so much with me. I am stopping now every ten minutes or so. I rest on my poles. Fifty-five pounds (twenty-five kilograms) at least on my back. Fatigue overwhelms my body. The straps are cutting into my shoulders, the pain worse on the shoulder which I have the gun on. Tired, I fall over on a ridge and it is a grueling task to lever myself up again. Beyond the first rubble field, the ice is chafed and creased before becoming corrugated, and then another rubble field. The last few miles have turned into a godless, frozen obstacle course. This was never meant to be an easy journey.

I try to push myself harder, breaking the last stretch down into sections measured by the position of icebergs. With the determination and anger that I am attacking this final stretch of sea ice, anybody would think there were gold on Herbert Island. Instead, there is permafrost, rock and half a dozen abandoned huts. The tiny unidentifiable specks barely visible on the horizon half an hour ago have become the houses which were once homes for a few hardy people living there. I champion a second rubble field, and make it finally to the shore ice. Having spent approximately three hours clambering over frozen pressure ridges the size of demolished office blocks, I am exhausted. I leave my skis and rucksack on the shore ice.

The snow here is deep like in Savissivik and I find myself disappearing into snow holes on the short walk into the settlement. The first house is a relatively modern red A-framed house with a ladder leaning against the side of the wall. The house is all locked up. Standing on tiptoes, I can just about see into the living room. The house was vacated in a hurry, leaving used cups and plates on the dusty kitchen table. The remaining houses are tiny, weathered cabins. Some have lost their roofs; all that remains of others are shaky wooden frames. Polar bear fur insulation in the cavities of the walls. Most houses are locked with a padlock you might use for a suitcase. Many of them look like they have not been lived in since the outpost was abandoned. There used to be a shop, a school and a church here. It was a functioning community.

I come across a derelict cabin which once may have been painted green. A piece of string is tied around the frame of the lock. I remove the string and enter into a porch full of rusting tools and frozen paraphernalia. Inside, there is a stained mattress on an old-fashioned sleeping platform (*i'dleq*), a ripped black leather armchair and a small, scuffed table cradling broken glass. Piled up books, a disconnected oil heater, empty grubby jerry cans and a discolored crucifix hanging lop-sided on the wall. Stiff, frozen coats hang from rusting hooks.

I decide that I will go and get my rucksack, change my clothes here, have a cup of tea and then work out if I might be able to sleep in one of the huts with the aim of exploring the island tomorrow. The temperature in the house must be the same as outside, minus twenty-two degrees. I have brought with me my sleeping bag and an Emergency Survival Heat Sheet. Trudging through the very thick snow, it is quite a long walk to get my bag and it is only then that I realize I have made a major mistake. I have not yet changed out of my sweaty base layers. I am wearing four very thin base layers with a tatty puffer jacket over the top and that is all. I knew it was going to be a long day and that I would be working really hard and getting hot.

By the time I have collected my bag from the shore ice and made it back to the abandoned hut, it is too late. The sweat has frozen and I am completely unable to get warm. I jump up and down, bang my feet together and do everything to get some heat into my body, but the pins and needles in my fingers are getting much worse: I am losing the feeling in my fingers. They may as well be solid lumps of ice. The sweat in the inner glove has frozen in a matter of minutes and I cannot warm up my fingers. I try to pour myself a cup of tea from the Thermos, but my frozen hands cannot get a grip on the lid.

I struggle to get the zip open on the back of the rucksack with frozen fingers. Inside are the heat warmers which are small packs of a granulated mixture of natural ingredients that when exposed to air react together to produce heat. I curse at the packet cover: "wait fifteen to thirty minutes for heat." I strip off in the cabin and replace the base layers, but it is no good as my body temperature is still falling. I just cannot get warm. Violent shaking. First stages of hypothermia.

With disabled hands, I try hopelessly to fold with much care my clothes into neat parcels before repacking the rucksack. The onset of mild hypothermia is making me confused, allowing me to forget the urgency of the moment. This is no time for tidiness and order. Then, a moment of sanity. I realize that there is only one option: head straight out and start skiing again. After many hours of carrying a heavy rucksack with a cumbersome rifle cutting into my shoulder and skiing more or less non-stop in temperatures of minus twenty-two degrees, this is the last thing I want to do. I desperately need something of a rest, but this will surely be the quickest way of increasing my body temperature.

I stumble out of the hut, stooping to open the rotten door and a honey-colored dog runs past me. Changing into dry clothes has given me some relief, but I am surely imagining this. The confusion is alarming. Then Ingaapaluk appears, wearing his polar bear skin trousers, seal skin *kamikker*, traditional anorak and carrying a whip in his hand. His twelve-piece dog team sits obediently behind him at the bottom of the slope, watching every move the hunter makes, waiting for the *qimmeerriut* ('dog that has broken his lead') to come back.

Nobody has lived on Herbert Island for over twenty years. This stronghold of freedom gave itself back to nature long ago. The occasional hunter comes here, but that is all. Here I am. I came here to supervise over this emptiness. After all I wanted the life that this space implies. I am a shivering wreck, tens of miles from anywhere in an abandoned settlement and yet there is no expression of surprise on Ingaapaluk's face. He asks me whether I skied here and where I am off to next. He offers me a lift on his sled. With chattering teeth and an unsteady voice, I tell him that I am cold, that I must go and will see him on the ice.

I make myself jog along the rocky shoreline covered in slippery ice. With frozen toes, I am stumbling all over the place, but I push myself as hard as I can and try and work up a sweat. I run as if my life depends on it. The heat warmers slowly give some warmth to my digits. The feeling is coming back to my fingertips. I clench my fists. The survival instinct kicks in and I push aggressively on my poles, pushing up the tempo and ski as fast as I can. I am relieved to have turned the corner.

After half an hour or so of intensive exercise, I hear the familiar sound of the panting of dogs behind me. There is a penetrating wind sitting on the sled, biting into me, finding every tiny chink in my armor. I chat briefly with Ingaapaluk. He has been checking seal nets near Herbert Island. Tears pour down his face from the punishing cold and freeze more or less instantly.

At the hunters' cabin with all of its blessings of remoteness, I defrost my fingers and toes slowly over the heater. Like toast on a grill. It will not get dark tonight, but I soon fall asleep.

April 16

Back in Savissivik, I browse through piles of tatty, black and white photographs of Qeqertarhhuaq (the name of the abandoned settlement on Herbert Island): eddies of memories that bring a tear to the face of the Ingaapaluk. It turns out that the house I went to was in fact Wally Herbert's and that Marius Qaerngaq, who was shot by his wife on the island was Ingaapaluk's father. They lived in one of the houses that has been smashed to pieces. Subsequently, his wife and killer took her own life.

Even after the drama of my trip to Herbert Island, the allure of the harsh, frozen world remains overwhelming and runs deep in my veins. The emptiness becomes obsessive. It is more than a physical setting; it is a state of mind with deep figurative significance. I enjoy the risk; the sense that anything can happen; the jeopardy of circumstance; the creeping up to the precipice; the inseparability of danger and opportunity. Danger is interesting and necessary for the human spirit.

But what does it mean to seek all these thrills? What happens when the appeal of the thrill begins to wane because you have tried to turn your life into a diet of thrills? C. S. Lewis thought this was the quickest route to becoming old and disillusioned.

April 18

The light is everywhere. The settlement feels warm. But on the sea ice, it is a different world. Still minus twenty-five degrees out there.

Morning devoted to scrutinizing satellite pictures of sea ice. Further adventures are germinating in my mind. I pore over the satellite imagery as a surgeon would scrutinize a CT scan. A collage of pancake ice, open water, polynyas and leads that differs from one day to the next. The ice is alive. The images show cracks, fissures, veins, capillaries and arteries of mushroom-colored ice. The constantly changing configurations are a confounding challenge for the polar traveller. In a few weeks' time, this oeuvre will be smashed up, redistributed and scattered. And some weeks after that, it will be as if ice had never been here.

In the afternoon, flags fly at half-mast. A teenage boy found hanging from the ceiling, noosed. Swinging to a premature dirge. Another tragic tale of unrequited love. The men talk about it in a matter-of-fact way. Hunters shrug their shoulders and gesture with their hands in a manner that suggests standing in the way of fate is futile. If God has predetermined everything, there is perhaps little motivation to understand the source of the problem.

April 20

The morning is spent trying to protect my solitude from visiting bachelors with little to say. I am seldom at my best in the morning. So much for me trying to find peace at the edge of the world. In the end, I find the best solution is to be outside drifting in the snow.

Afternoon: observing arctic hares on the tundra. Flashes of white on white. They dart around at incredible speeds. They play noughts and crosses in the fresh snow leaving the most indelible patterns for me to photograph.

With the changing climate, food has become scarce for the hares. In recent years, there has been unusual rainfall in November and December. A thin layer of ice over the lichen prevents them from getting at their food. Two years ago, many of them died of starvation. The movement of the hares appears to have changed too with them travelling further and further to find a mate.

April 22

A life less shared. I am free from all the platitudes and empty talk that become such a mental drain. The only laws here are my own. Well, until the bachelors come visiting.

April 23

The night was full of dreams moving through my sleep like schools of fish. The sleep mixed up my various identities. Morning. New snowfall. Heavy, tired snowflakes. Winter has returned from somewhere. Will it ever end?

Late afternoon. Drinking tea. Ibbi comes to visit. Funny stories pour out of his mouth. Hours of ribaldry. A never-ending circus of smiles and banter. For example: Lauge Koch, the Danish geologist, was here drawing maps in the 1920s, and he and the Inuit went up to Cape Alexander. Not far from the point, Koch spotted an island and asked the local what was the name of it. They told him the local name which translates as 'erection'. It was a pointed island. Koch saw no reason to question it and so it appeared in a map published in the 1920s as 'erection island' *tiggarniq*.

Before he leaves, I ask him: 'Ibbi, why did you choose to stay here?' (and not move to the town).

Ibbi: *ihumainnahorjamahunga* ('I want my freedom', literally: 'I want my mind to be open to thoughts'. He nods seriously, looks out across the sea ice and then adds: *ihumainnahortortoqainnga* 'I always want my freedom.'

No explanation required. I understand precisely what he means.

And, then he is gone.

Note: my biggest concern regarding potentially writing a book about my experiences in the Arctic is that more people will come here. An oddity about cabin life and life on the sea ice is that somehow you want all the beauty for yourself. In this respect, the Arctic is perhaps like a beautiful woman.

April 24

Imagine a place without cars,
Trees, rain, thunder or grass,
A place where lost thoughts are awoken
By the smell of pungent *kiviat*

Legs wide apart, a family of four sit on the kitchen floor feasting on long dead little auks (*kiviat*). Sleeves rolled up, they bite into the red flesh. Smiling and grinning. An elderly woman hacks at a piece of frozen caribou meat with an axe. A *kaffimik* is being held at Jens Ole's cabin. His grandson, Iihaq, turns six today. The birds' wings are removed first and then the feathers are plucked. All the organs are eaten. Our arms and wrists are stained in blood. The taste: something between pheasant and Stinking Bishop. Dinner finished. We look down at our small piles of bones and say *qujan* ('thank you'). Not to our hosts, but to the remains of the birds for 'giving themselves up to us.'

Thought this evening: the Inuit have no problem rationalizing their relationship to animals. In the West, these issues have risen prominently to consciousness. This must in part be because factory farming has removed us so far from the act of killing what we eat (we like to pretend things never lived), but it might also reflect the decline in religious belief. Previously, man knew his status as the highest order of creation. Made in God's image.

April 25

Good Friday. I make it to the church. About half of the community is there. The children cough, snort, sniff and hawk up phlegm throughout the service. Oblivious to the hermeneutic of the congregation, the lay-preacher has a sad face and speaks without confidence. Nobody cares. Afterwards, the congregation rushes outside and lights up cigarettes. I am reminded of Geoffrey Hill's *Canticle for Good Friday* told from the point of view of the Apostle 'Doubting' Thomas. All that meaty physicality of language. It could only be Hill with talk of "carrion sustenance" in the "unaccountable darkness."

Enough canticles and talk of resurrection, I go outside and look up at the sky. Heaven. The sky blue; the ice whiter than I ever imagined. The clarity of everything seems conspiratorial; the view almost too perfect to be real. All fixed and glassy. On such a day, dying and going to heaven wouldn't be worth it. Untainted shadows spread across the fjord like water spilt on a white carpet. The carpet is stretched taut to the points of the land. The scene looks like a still life, but it isn't. This place is fluid. Down by the shore, last week's full moon has broken up some of the perfection. The ebb and flow continues under the ice. Large, deep cracks are now appearing in the tidal zone. The bears have retreated with the ice, north to the Kane Basin. Skinned arctic foxes hang up outside hunters' cabins. Seals are beginning to emerge from their breathing holes. Living it up and basking in the sun. I can't blame them.

It is getting warmer with each day that passes. For the first time in months, I have discarded the trouser base layer and am now just down to three layers on the top (minus coat). Soon, it will be spring, right?

Thought this evening: Why do our words for 'remoteness' carry such negative connotations? Such sinister associations. You can be remote from the world of men, but close to nature. Then, it is not so much 'remoteness', but a kind of divine intimacy.

A dog team appears as a tiny speck on the frozen Murchison Sound

April 27

Morning. Radio fuzz blows into the kitchen. A procession of nightmares last night punctuated by the cries of Lenin. Ibbi insists it is because I am living in an *i'ddu ajorpoq* ('bad house'). He thinks that the house is haunted. No problems sleeping though. I fell asleep to Bartok's piano concertos last night! I didn't know that was possible.

Twenty-four-hour sunshine. I do not tire of the endless light. I celebrate the Easter festival with Edmund Spenser:

> Most glorious Lord of lyfe that on this day.
> Didst make thy triumph over death and sin:
> And having harrowd hell didst bring away

Captivity thence captive to us win:
This joyous day, deare Lord with joy begin,
And grant that we for whom thou diddest dye
Being with thy deare blood cene washt from sin,
May live for ever in felicity

(*Sonnet LXVIII*)

April 28

A small meeting has been held about my mattress. I was nearly hypothermic in my own cabin for patches of the winter, and nobody cared. But as soon as Ibbi mentioned nightmares and the haunted house, there is deep concern. I am asked to join the meeting:

'Have you seen any ghosts?', one committee member asks me.

'No', I respond.

'Anybody knocking at the door late at night?', he continues.

'Louise Simigaq', I hear one committee member whisper. Hushed voices soon circulate the room. Much serious nodding.

The consensus is that the mattress is the problem. There is only one dissenter, Jens Ole. He thinks I have nightmares because I do not eat enough meat. He points to the veins in my hands: 'bad blood circulation. More seal meat.' Louise Simigaq, a visitor, slept on the mattress before and had nightmares every night. In the night, she would hear a knocking at the door. She would go to the door and there would be nobody there.

Late afternoon: Ibbi and some of the Committee members come round with a new mattress that has somehow appeared *ex nihilo*. The 'cursed' mattress is removed, but not before the young children jump up and down on it. They scream and run away, frightened that any evil spirits might pursue them.

April 29

Minus thirteen degrees overnight. A full fifteen degrees warmer than just a few days ago. I turn the heater off. I slept well on the new mattress. No nightmares. Problem solved.

Evening: Shadows run across the ice. I am out walking. Reason: I need some fresh air, much needed exercise and time to digest the aphorisms and

Lebensphilosophie circling in my head. Other purpose: to sit and watch the seals basking. As camouflage, I have with a me a *tarhaq* ('a white shooting screen that hunters lie behind when taking aim at a seal'). I follow in the tracks of arctic foxes. I lie head down on the ice behind the screen. I cup my ear and place it on the ice. A wind from the west. A ghostly hum. The ice sings. The binoculars are glued to my eyes. I scan the horizon laid bare before me. In the far distance, a spot of grey. A ringed seal perhaps. I move closer, but then it is gone. The procedure is repeated many times. And so, I return to listening to the ice. The pressure ridges crack and groan. I roll around in all the seismic bliss, and return to the cabin feeling thoroughly refreshed. It is these finely calibrated shifts of feeling that make us human.

Thought back in the cabin: most of us assume our lives could be depicted graphically as a kind of symmetrical bell-shaped curve. We assume that we peak physically at least in middle age, and then slowly our bodies and even minds degenerate in some way. And yet, if you are a true 'believer', your chart might be quite different. The line rises indefinitely at forty-five degrees for you are in pursuit of Heaven or Paradise, and when you reach it, you just continue to scale the heights.

May 1

A few more new words:

> *qa'pvaq*: 'a big area of multi-year ice with clear white blue colors that can quickly move into the area and block off the shoreline'
>
> *napar'ddilugginnaqtoq* 'somebody who just hangs around, lacking the will to provide for himself'
>
> *ikiarhihutin*? 'have you lost the plot?'

May 2

I wake up to what I think is the friendly chirp of a snow bunting. Has spring arrived? They have more trust in the coming of spring than me. Excited, I go and investigate. Young dogs in pursuit. Outside, life is going about the business of getting on with itself. I squint against the shards of sunlight jabbing at me. Perched on one of the store houses, the cheery little bird in his black and white breeding pajamas chatters away. A smile stretches across my face. I take his greeting personally. More or less convinced that it is meant for me alone. I wondered if he would ever return. Their migrations make my travels look like small fry. He is the first to return and is making claim to the best real estate. He will not be alone. They never are. Soon, the Arctic will be brimming with life.

This pioneer hops around in front of me. I know its curiosity. It is my observations and constant curiosity that make me. I prefer to carry questions through life rather than arguments. The mind should never be satisfied. All the time I can remain curious, I know that I can hold onto something that we all have as children. Surely, we are only as great as our capacity to wonder. My mood is now perfectly attuned to this pioneering bird. It is almost impossible to resist the anthropomorphizing impulse on such an encounter. So, the cycle of seasons has finally made it all the way up here. The birds are returning. The Earth is still rotating. Life is truly starting again.

Down by the shore, dogs feast on their own excrement. Entwined in their offspring, young boys unfurl long whips. *Aulaitsit, aulaitsit.* The hunting culture is not going to die out just yet. Nature might still win for a bit longer.

Ibbi visits in the afternoon. He is talking apocalypse. One of his favorite subjects. In the event of a major world disaster, he believes the Inuit would be the only people in the world who would be able to continue to live because they have not forgotten the skills of their ancestors. Spend enough time up here, and you end up subscribing to the narratives of exceptionalism. That is the way we build strong ties. We are not the *hoi polloi.*

Memo: the solace and delight I find from the birds (and animals), absent for so long, is almost inexpressible. For he who lives alone in a cabin, these are the greatest companions. And yet, hunters tell me there are far fewer buntings than there used to be. Such stories are familiar to people all over the world. The dinosaurs failed to anticipate the meteorite that extinguished them. Are we going to make a similar mistake?

May 4

At the shop, I ask Taateraaq which day it is. *Pingahunngorniq* ('Wednesday'), he responds. And, so I decide to try and access the Internet via the satellite to see what the news is. My love has left me. She wrote the e-mail a week ago, but I have only seen it now. She doesn't want some kind of Denys Finch Hatton character as a boyfriend. Always craving his freedom and looking for new adventures. I cannot blame her, but nonetheless feel a little shocked.

It was absurd to ask her to wait for a year whilst I lived like a hermit in the High Arctic. I feel no anger. I understand that idealism scores few points. I feel only remorse that I was unable to be a more dutiful boyfriend. I got what I deserved undoubtedly. She is doing the right thing, but I still feel the pain:

> And that taxing courtship in a far-away land,
> Whose complex narrative now lies silently in the snow and sand,
> Like one of those epic films that we love so much,
> We almost made it, *mein Liebling*,
> It was always just over the brow of a hill,
> Just over the brow of a hill

May 5

I spend the day walking along the shore line. My mini dog team is in tow. Do they know anything about sorrow? I want to walk as far as I can. I want to tire myself out completely so that I am no longer able to think about lost love. It seems like there is not a soul to share my grief with, so I fire questions at the dogs. In times of despair, you need a dog. In times of *anything*, you need a dog in fact. The madness and selfishness of this adventure begins to creep into my veins. Here I am in the Middle of Nowhere. But I am not just passing through. I am here for a year!

I return to the cabin and dip into some D. H. Lawrence. He wasn't fussed about dying at the age of forty-four. He had apparently done and seen it all:

> I should find, for a reprimand
> To my gaiety, a few long grey hairs
> On the breast of my coat; and one by one
> I watched them float up the dark chimney
>
> (*Sorrow*)

On love: love is the product of a certain kind of spiritual affinity. It is something that comes about spontaneously, and can be triggered by quite tiny details. You can fall in love watching the way somebody eats. Equally, such small details can be a source of revulsion. Love is not something that slowly emerges over time. That might be respect, friendship or any number of other things.

At heart, most of us are in fact optimists. We make mistakes in love time and time again. Feelings come and go after all. Distinct from love, 'being in love' usually does not last. We all know that, but like to pretend otherwise. We know how cruel that can be, but we are convinced we will be the exception. Our hearts get broken; we break others' hearts. But we carry on in this life convinced always that we will end up finding the right person, an eternal and (almost) perfect love. And for that to happen we should perhaps negotiate tenderly the transition from 'being in love' to love.

May 6

I slept badly, wrestling with the memories of a love story and the lost fruits of passion. The morning portends more warmth. Minus fifteen degrees. I want to be out of the cabin, no longer wishing to stare at four walls asking searching questions, challenging assumptions. The need to strike up a conversation. So, I visit one of the bachelors (a kindred spirit?) in the hope that he may lift my mood. Aleqatsiaq. I remember meeting him right at the beginning of this adventure. The locals call him the 'wild one' (*nujurataq*). He moved here for the 'quiet life' and to escape the endless sun of the exposed settlement called Qeqertat. He occasionally visits the odd relative that lives here, but tends to keep himself to himself.

The scene: his prefab cabin is basically a pigsty. The walls are caked in grease and grime. The sticky, blood-stained floor a dumping ground for beer cans. Dressed in filthy black salopettes and clogs, he smokes unfiltered cigarettes. One after the other. He clearly couldn't give a shit. The tiny cabin is intolerably smoky, but I want to stay a while. Cigarette hanging out of mouth, he grunts something, breaks wind audibly and puts a saucepan of water over the paraffin heater and prepares some coffee. No kitchen. No electricity in Aleqatsiaq's cabin either. The electricity unit blew up in the storm of 2007. He grins, and seems quite happy with the status quo. No impulse to 'modernize' here. He will never become a materialist. No thrill-seeker, he is perhaps seeking another level of simplicity. At any rate, I decide not to raise the question of why he never married. I think I have all the answers. Instead:

'How do you manage in the dark period without electricity?', I ask.

A shrug of the shoulders. A finger stub points to rivers of dried candle wax that decorate the table. He talks about the hunters, the old days, the little auks that are expected any day now. His creased forehead is a map of nostalgia. He seems to crave an earlier life.

Whilst talking, he plays the cat's cradle string game. An exceptional cat cradler. Intricate figures produced in a flash. With one piece of string wrapped around his hands, he makes the shape of a walrus, dog, seal, seagull, anus, vagina and penis. The list of achievements comes as no surprise. He shouts out the names of genitalia. His face explodes into laughter.

I sip my coffee. He gollops down saucepans of caffeine into confusion. We chat for a while. A short silence. A twitching of eyebrows. Then, this. Aleqatsiaq stands up and pulls his pants down, urinating into an empty fruit juice container. No bathroom here. He then flicks on the radio. It is three o'clock and he wants to listen to the news: thirty-six suicides in Greenland so far this year. My mood sinks further.

The news finished, I take my leave and walk to the store. The shelves are now more or less empty and will not be refilled until the supply ship comes in July. I catch my own fish, and buy meat from the hunters. As for fruit and vegetables, no sign of those. I suck on blubber for Vitamin C.

Hiuliqatoq: 'a person who likes to live like in olden times'

Memo: what do I *not* like about cabin life?

- Unwanted visitors
- Being besieged by lust
- Oddly enough, the lack of privacy
- The thought of having to explain all this to a hard-and-fast urbanite

May 7

Chi-tik, chi-tik, chi-tik. Two snow buntings indulge in vernal foreplay on the balcony. There is a background chorus of dogs howling. It is probably time to get up. If you live in a big city, your dawn chorus is muffled car horns. Here, it is puppies and snow buntings.

Last night it was actually too hot in the cabin. I never thought I would write such words! It is minus four degrees outside. The morning delivers a dusting of snow. I have no idea which season it is, but everything is in flux. Migrations; summer arrivals; the shifting of ice floes. It is meant to be spring, but I am still waiting for the binaries to dissolve.

A screech of gulls sits on the sea ice in grey and white plumage, congregating and making plans, looking cold, lost and ill-at-ease. Announcing their arrival, they work their way through a melody of bugle calls starting with the *Reveille*. After months of silence and no birdlife at all, except for the occasional raven, the Arctic has suddenly come to life. The first few little auks have arrived by the cliffs behind me. They are five days early. Things are changing. Soon the sky will be black with migratory flocks of these laughing birds as they arrive in their millions to their breeding grounds. The hunters tell me the dovekies fly about 2,000 miles from the North Atlantic to get here. Without the stars of the night sky to guide them, they come to Savissivik the same week in May every year and then leave again on August 7 or 8. Despite the constant daylight, not only are their circadian rhythms intact, but they even know which day it is!

Thought this evening: living in the tundra risks some kind of mental trap. Here there is time and space to think, cogitate, scrutinize, reflect upon and think again, but I fear the impact and evolution of my thoughts is somehow trapped in this environment. They may have no currency elsewhere. They are only meaningful to me. Utterly private. They belong in the High Arctic, and 'down there' such as in a crowded city, may simply be met with a bemused stare.

I think I have retreated far enough into this inner world, producing fantasy dialogues with the dogs and even covert, invisible lovers.

May 8

The siege is underway. Lines of bombers pour over the cabin at great speeds. The air smells of burning peat; the sky thick with the laughter of small tuxedoed birds. This is an important day for the community. Hunters have been getting their nets ready and talking about nothing else. Spry men will dangle over the cliff edge and pluck them out of the sky with their nets attached to great long poles (*ipoq*). As the birds fly over, the wrist is twisted. Once the bird is caught, the wings are rotated so that it cannot fly away, and then the heart is pressed, the fingers slowly pushing up towards its neck. Internal bleeding and the bird soon dies.

The hunters will catch hundreds of these birds, and 'ferment' them in a seal skin with blubber which is placed under rocks. Six months later, they will eat the birds. When the little auks come, so too do the falcons and the arctic foxes. The foxes—these hardy, scavenging nomads—sit at the bottom of the cliffs, dodging the fecal fallout and waiting to pilfer any eggs or fledglings that fall off the edge. The falcons soar above the feast. In turn, they will be mobbed by ravens doing their piratical thing. This mini eco-system is going to soon kick into action: the lichen needs the guano, the arctic hares need the lichen, and so on.

I am happy to sit out the drama of catching seabirds and watch it from afar. Stopping thumping hearts is not really my thing, and there has been enough heartbreak recently. But this is sustainable living *in extremis*. This is not 'consume and throw'.

Instead of netting dovekies, I go down by the shore with my binoculars to inspect particularities. In awe of the painting before me. In the non-stop light, I prefer to sit outside. Sunlight blazes down on me. Waders scream past me. Calligraphic leads are beginning to be inscribed into the sea ice. If it were not for these changes, I may no longer be sure that time is still passing. The sun sits high in the sky. I pore over maps with fierce concentration. There always has to be *another* trip.

My acoustic space is now entirely different: coalitions of auks and shrieking gulls squabble and bicker like parliamentarians. Continual skirmishes. Raucous courtships that border on energetic bankruptcy. The silence of winter has turned into the interminable frenzy of a major breeding ground. Nature has exploded. Testosterone pumps through the blood stream.

* * *

Deliberation this evening: I have had just occasional, patchy Internet access for nine months. How has life been without those couple of hundred checks a day of the smartphone? Like any drug, initially it was difficult. Now, life feels so much more *real*. When you are not living in the moment, your holistic vision shines brighter. Conversations replace Google searches. Emotions and more complex critical thinking replace memes and hashtags. Poetic aphorisms replace Tweets. Tearful valedictions replace e-mails. Gossip replaces news items from distant places. We are so caught up with being up to the moment we have little sense of what stands outside of time. When we step aside time, we care less for being told constantly what has just happened but wonder about what has just

not happened. Another thing. Now, I remember
if I had almost forgotten what a human voice so
more acute. I feel more creative. It may seem li
but I feel somehow as if my sensitivity to the
profound. These are just some of the effects of n
ter over us, and in shifting from virtual reality t
to go forward and *progress*, I have had to go b

What is creativity? Discovery and imagination.

May 9

Some highlights from the local diet:

January: caribou

February: walrus

March: polar bear; arctic hare; ptarmigan

April: halibut; arctic char; polar cod; arctic fox

May: seal

June: eider ducks; guillemots' eggs

July: narwhal

August: crowberries; bilberries

September: dried narwhal skin

October: fermented Greenlandic shark

November: fermented little auks

December: musk oxen

Vegans: this place is not for you!

May 11

I spend the day making a small sled for the children. It is basically a glorified palette on a set of old skis with some wooden uprights at the back. Four hours of work and by mid-afternoon, we have our sled. The children will have a dog sled race this week. I am not confident that my contraption will end up yielding any trophies, but I am surprised how much I enjoyed

…gether. My mind was occupied completely on the task at hand. I … a goal. My total lack of skill in such matters resulted in something …ble. And that is sufficient for one day.

Memo: a few more thoughts on technology. Once our lives are completely technologized, we will care even less for words. Fifty years ago, grammar was the guardian of thinking. We cared about how we spoke and what our thoughts, once written down, looked like. I am neither a prescriptivist nor a stickler, but if we let language be debased by technology, we will have deprioritized something precious which makes us human. Technology will spurt out the 'truth', and we might lack the rhetoric to challenge it. Language should allow us to stay true to the human experience, and be a means to draw on a deeper place. If it does the opposite, we risk ending up in some kind of Orwellian nightmare.

May 13

Morning. The dogs are howling. A hunter must have returned to the settlement. More seals have been 'sacrificing themselves'. I am up. Cirrus cloud cover reflected on the sea ice. The sea ice looks like a chess board. A handful of ivory gulls have arrived; almost entirely white, they hop from one square to another. My attention to each object and activity has become almost sacred. I have become a "monk of this world", and the white desert before me is the metaphor for that inner awareness.

And, so I am out on the ice again. This time I am *skijoring*. That is to say I am being pulled by *Nipi* and *Nukka* on skis. I want to share all this beauty with them. The dogs are pulling hard. The ice is for the most part smooth. I am moving fast, more graceful than ever before on skis. I slow down so that my vision is appropriate for the landscape. I am heading west, towards the ice edge. A skirl of kittiwakes overhead breaks the silence reminding me that we are not alone. What am I looking for out here? Am I trying to be just something more than a banal human being? This is no self-discovery mission. Haven't I had enough solitude and silence? Adventure, risk and freedom, yes. Always. But what else? Gaston Bachelard said that "immensity is within ourselves." Alone, we come to these places to interiorize the desert and so feel small again, and let that immensity loose.

I feel like a *Desert Father*, one of those fourth century spiritual seekers that left the material world behind and lived a monastic life in the desert. The holier they became, the more dependent they became on God's mercy.

Unlike the *Desert Fathers* though, I am perhaps not looking for God and would make for a shabby renunciant. I can make do with another sensibility and profound wonder. This is where philosophy begins and ends, isn't it? My mind wants to leave behind the minutiae of modern life. The polar desert is a metaphor for the idealist's escape from modernity, for the traveler eager not to have his perceptions blunted by materialism. I am somehow convinced that there is more meaning out here or at least that this harsh, raw beauty will strip you bare of all your pretensions so that you can at least see what is meaningful. So many of us are looking to break free; so many of us are becoming spiritually at least *Desert Fathers and Mothers*.

The temperature drops sharply in the evening. Winter's last push. I try lighting the heater, but put too much oil in. The flame shoots up, burning my right hand and my hair. The oil heater is getting its revenge for all the kicks I gave it over the winter. He wins. I look like George Brewster.

Life is laughing at me again.

Nipi

May 15

Insomnia. With the constant light, it has become very difficult to sleep. A prisoner to light. I lay awake for hours, tired but unable to sleep. I listened to the oil heater snoring. The sunlight has given me so much energy, I do not feel tired until the early hours of the morning. I could not even fall asleep to Bruckner last night.

The smell of burnt hair lingers in the cabin. Rotten eggs. Not even the bachelors will visit. The stench is another good incentive to get up early and out of the hut. Savissivik is hidden in dense *pujoq* ('smoke', 'fog'). The mist expunges my sea ice. Behind the cabin, I can hear the sound of running water. The temperature has risen. Meltwater is pouring off the slopes and picking up dirt and mud on the way as it leads onto the sea ice. The puddly ground squelches under me. I meet Taateraaq. He walks metronomically down by the shore. He stands silently, shuffling his thoughts. Clouds scud across the sky. A mood of loneliness travels across his face. He points to the north beyond the cliffs where the birds have settled. 'Tomorrow comes the *avangnaq*', he says. 'Some of the ice will be blown away.' He grins and then is on his way.

May 16

Gusty weather today. As predicted. Northern winds thrash Savissivik until noon. But then calm. The view has changed; everything is foreign again. Icebergs collapsed into piles of rubble. All my paths and tracks folded together; the sea ice—a palimpsest of time visible. Open water in places. Flustered gulls reworking their plans. Lots of mewling and swooping. The borders of my desert, that harsh playground, are unrecognizable. The parameters of my freedom have been clipped.

* * *

I write this diary each evening. Images, bits of conversation, ideas. Always sat by the window. Sometimes buffeted by temptations. Often with a cup of steaming tea in my hand. I have this fear though. I am trying to archive all these past hours as if each were packed with significance, but I am worried that there will still be days when I simply cannot think of anything to write at all. What will all this scribbling amount to? What if my internal dialogue were to dry up entirely? Then, I imagine how my diary would look if I wrote one when back at home. Surely, almost each day would be filled with the most banal, mundane entries. I would be left describing my dinner and whether the pasta is sufficiently *al dente* whereas here I can look out the window and see the pantomime of crashing ice floes, exploding icebergs, skeins of snow geese flying over in their energy efficient V-formations and listen to the little auks ('penguins of the North') ridiculing it all and shouting 'he is behind you.'

May 17

The dogs are quiet. The silence of the tundra penetrates the cabin.

May 18

Savfak, a kind elderly woman, is telling me stories whilst she spools her tapes back into cassettes with a pencil eraser. She is complaining of headaches. Her explanation for this makes for a remarkable account.

In 1968, a B-52 bomber belonging to the US military landed in the sea not far from here. There were four nuclear bombs on board. One still lies on the seabed. The Americans covered up the accident for over thirty years. The explosion was of such a magnitude that an enormous ball of fire could be seen hundreds of miles away. It happened during the 'dark period' and the Americans had to bring back the families that they relocated as they were the only people who knew the region well enough. She knew the hunters that worked in the clean-up operation. Lives were lost. The food tasted bad after the accident and many people suffered with various illnesses. She shows me photographs of deformed sea mammals and points to the places on the maps where they were found. She sheds a few tears, but then quite suddenly is as right as rain again.

This is not the first time I have heard such stories. A small community of remote Arctic hunters could have been wiped off the map by the greatest military power on the planet, and we would have never got to hear about it.

Memo: these David and Goliath encounters play out all over the world and are relevant to all of us. Indigenous groups in the Amazon Rainforest are witnessing their habitats being destroyed by multinational logging companies who shoot to kill anybody that gets in their way; closer to home, mining companies are savaging the ancient herding grounds of the Sámi in Scandinavia. These accounts are allegorical. It is the same for he or she who is the lone voice of reason when faced with corporate or academic groupthink. But it is better to be the lone voice than to seek the 'security' of the Nietzschean lowlands of life.

May 20

I am lording it over the ice. Avoiding all routine. It is a warm day. Feels almost like summer. The ice is flat and I am practically skating across it on my skis. I feel a strange sense of responsibility to survey constantly all this beauty, to record its changes and tell a few stories along the way. All this nature can teach us something profound. Some find their transcendence in works of art, but it seems to me the wild places left on our planet is also not a bad place to start. They remind me of what I am living for. We all have these inner teleologies. We are all striving towards the Self, but the path can be torturous for our lives are so busy and complicated.

* * *

Far from the settlement, men and women are now fishing in open leads between five and ten feet wide. They tell stories; their narratives wed together their kin and that is probably their purpose. The leads reveal the life underneath, hidden from us for so long. The ice is retreating. For the moment, I can ski around the leads but in a week's time, I suspect it will be impossible. And, so my mind is made up: I want to do another long trip before all this ice is swept away. Afterall, that is the journey that kept me awake last night. I want to pass through all those blank spaces left on the map. I'll see if I can ski up to the northern most settlement, Siorapaluk. I borrow a pulk, satellite phone, Thermos and Primus and begin to make preparations. This time, I will leave my mini dog team at home. The sled is heavy and I will be more flexible with a light-weight pulk.

In the afternoon, I pack the pulk up and have a dummy run on the ice. Everything runs smoothly. Boyhood dreams flash across my mind. This is the life that I wanted. I cannot allow myself to fall asleep to the true nature of life.

This evening, for the first time I sleep with a blindfold. The thin pink curtains are unable to prevent the sunshine breaching my exposed little cabin twenty-four hours a day. My sleep is becoming increasingly disturbed because of the constant light, but it matters little. The sun gives me energy; my mind busy concocting new (and no doubt unfeasible) adventures. For the first time in my life, I have become a poor go-to-sleeper, having trouble parting with my consciousness.

May 21

It is a Sisyphean task carrying two rucksacks, skis, a pulk and a gun down to the shore. Dried halibut hangs on lines. Pools of water on the shore ice. Curious hunters joke with me, and wish to know where I am off to now. It takes me a while to ferry my equipment onto the thicker ice. All eyes are on me again.

It was mild overnight, and everything has changed once more. The sea ice has become slushy in places. A new lead has opened up. It is not a problem getting across it, but I can see that it soon will be. Have I left it too late? Let's find out.

After a couple of hours of skiing, a soufflé of melancholic looking cloud coming from the east sweeps across the bay concealing the almost melted down statues of ice. In a short period of time, it has gone from being almost clear to quite thick fog. The relationship between ground and sky is effaced. This does not look good. After a short while, I am no longer able to clearly see land and risk getting lost in such a self-similar terrain. Without too much hesitation, I decide to abandon the trip. I am not here to fight with nature, but to honor it. I could find my way up to Siorapaluk but this is not going to be any fun. If leads are opening up, visibility is quite important. If I can't see where I am going, I might fall into the icy water. I turn around and head back to the don't blink-or-you'll-miss-it outpost, pleased that I made the right call. Cloud and mist plummet down on me. A small crowd of hunters watch my reappearance through the fog. They could see this was a mistake, but were happy for me to find out for myself. They don't like to mess with your pride.

> *Ingen skamm å snu*—'there is no shame in turning back'
> (Norwegian saying)

May 23

The constant light is making me feel disorientated; a sort of metaphysical unease. Never mind which day of the week it is, sometimes I have to check which month it is. I go to bed at one o'clock in the morning, not feeling particularly tired but do not fall asleep until three o'clock. I set the alarm for 7.30am. Just a few strands of cirrus cloud garnish the sky. The sun is shining, but there is a cold, stiff wind. The sun appears from behind the ice sheet at about six o'clock in the morning now, and does not disappear until sometime after midnight. There is still thirty miles of sea ice. I am constantly told it will all soon be gone, but somehow it seems impossible to me. Once it is gone, I promise to shut up talking about said topic.

May 25

The sun casts long morning shadows. I am double-poling it across the ice. The rubble patches have turned into a sheet of glass. I have never skied at these speeds before. I am out exploring the extremities of the fjord before all this melts. Out on the ice, I meet Ingaapaluk. He is looking for *uuttoq* ('seals or walruses basking on the ice in the sun'). 'Very few seals in the bay this year'. 'I do not know why', he says whilst shrugging his shoulders.

Then, I hit a strong westerly wind blowing right down the middle of the sound. This spume of wind created by thermal pressures buffets me from all sides. With the wind, the temperature plummets and it is quite hard going. A half-baked attempt at a snow flurry. Visibility impaired. I am still learning the discipline of the Arctic. Suddenly, I slip on the ice. I land on my right pole. I look around and know immediately that the trip is over. I have bent the pole in half. I try to straighten it out, but the damage is done and it just splits in two. My fortune changed in a flash. Life is like that. Savissivik, this place that will always stand out in my memory, is barely visible in the distance. I am about thirty miles from home and now I only have one pole. I turn round straight away and head back. I have visions of me taking hours and hours to get back to the cabin, but nothing could be further from the truth.

I use the pole as a sort of paddle, taking strokes on either side and propelling myself across the ice. The ice is so smooth that this proves to be very effective. I am going nearly as fast I was with two poles. I cover the distance in just under two and half hours.

May 26

An Australian journalist is visiting Savissivik. She is doing a 'whistle-stop tour of the Arctic'. To me, it sounds like a bad joke. There is a knock on the door of my cabin. I know immediately it must be her. Nobody knocks here. I shout *agkerit* 'come in', but nothing. I let her in. She breaks into 'Oh, I am so sorry to disturb you. Might you be Dr Leonard?' kind of talk. Her way of speaking seems entirely alien. Here is the conversation (more or less):

Journalist: 'Don't you ever get bored here?'

Me: 'No, not at all.'

Journalist: 'Are you not lonely?'

Me: 'No, not really.'

Journalist: 'Oh, you have a television at least. That must have been welcome in the winter months.'

Me: 'Actually, I hardly ever watch it.'

Journalist: 'How do you cope without shops?'

Me: 'Well, I catch my own fish, and buy meat from the hunters. There is a small store. I buy my condensed milk there.'

After about twenty minutes of this kind of talk, I just want her to leave. It is obvious she is never going to understand what this is all about.

May 28

A pool of meltwater on the sea ice (*imatsinaq*) grows in size by the minute. Not that I am watching it obsessively. This meltwater straight from the ice cap is surely amongst the cleanest drinking water on the planet and yet it runs straight into the sea. We could be bottling it up and selling it to the Chinese. There's a thought for a Saturday morning. Well, I think it is a Saturday.

What is the news? The narwhals—those mythical submariners—have arrived at the ice edge. A few hunters have gone with a group of biologists to put transmitters on these unicorns of the sea and study their movements. Remarkably powerful dog teams now pull not just the hunters' sleds, but sleds bulging with equipment and visiting scientists that in turn tow motorboats with kayaks tied to the side. It does not look like it should be possible, but somehow it is. These four-legged haulers seem capable of anything.

An explosion of life. Visitors are returning from the autumnal castigation. I capitulate to anthropomorphism. Red knots: a medium sized wader with orange-brown jerseys trumpet their arrival. One, two, a handful and then a whole flock. They have flown 10,000 miles from Tierra del Fuego where the temperature is the same as here. About two degrees. If I were them, I would have stayed put. A few stop-overs in Brazil, Delaware and Hudson Bay and now here in the High Arctic for a summer holiday. The knots no longer crowd the beaches like they used to. Their numbers are down nearly 75 percent over the last thirty years. Preferring a smaller crowd, little ringed plovers with distinctive yellow eye rings dart along the shore, banking from one side to the other like TOPGUN instructors. Unlike the knots, they have been living it up in India. Above me, just having recovered from Louisiana's Mardi Gras, half a dozen snow geese (*kanguq*) in honking V-formations assess the breeding ground options. Thousands of birds have flown here from different corners of the world all apparently equipped with the knowledge that the sea ice is about to melt and reveal a feast of anthropods.

May 30

Everything is melting. I am out on the tundra. Returned to birdsong. Not wishing to miss for a moment any of the cosmic transitions. Flocks of Lapland buntings (*narharmiutaq*) peck at the ground amongst the cotton grass, blooming arctic poppies and the roots of the arctic willow. Overnight, we appear to have practically jumped a season. Shrubs, grasses, mosses, lichens.

Afternoon spent doing some wood-carving. I am making a replacement ski pole. Well, that is what I am going to call it. Job done, I hopscotch across the remains of the shore ice and back onto the thinning sea ice. Skiing for hours around the fjord without a care in the world.

Late afternoon. Fresh snow—the heart of innocence. Summer over? I am back in the cabin. The snow will conceal the grey patches of thin sea ice and make skiing dangerous. I turn the radio on: a polar bear has been shot not far from Savissivik. Oh. We thought they had all retreated to the Kane Basin. I might need that rifle yet.

June 2

May has become June. I am awoken by a gang of Lapland buntings having a ding-dong on the balcony. Morning spent writing poetry:

> I lie amongst the herbage of June,
> During the husks of the day,
> With imbalanced circadian rhythm,
> Under a diaphanous light,
> A head full of aphorisms in this introspective place,
> Drowned in my infinitely circular thoughts
> About my life, about our rented existence,
> About the pith of the matter,
> And the fundamental *Ur*-way of being,
> Thoughts that lead nowhere,
> But to the beginning, the beginning

June 4

The sunlight is incessant. I miss the evening sunset fiesta, a sense of day and night. Some kind of equilibrium. Despite wearing *two* blindfolds, I am awake much of the night wondering how such incandescence can possibly be conducive to sleep. I close my eyes and I see the lamp of my soul burning in a sleepless delirium. No dusk, no dawn, no escape from the endless light. The sun plays hide-and-seek behind the ice cap for a couple of hours but by 5.00am has reappeared on the other side to torment me.

I am up early for sleep is impossible. I drink one coffee after another and busy myself with errands. Ane-Sofie helps me clean the cabin. Then, wrapped in sunglasses, I am back on the tangled heath of the tundra amongst the sedges and wild flowers. I don't wish to be exiled from nature's sensuality for long. The snow has melted. The sky is black with careering silhouettes. The little auks are *en fête*. Fingers of elastic sea ice bend in the wind. Horse-shoe shaped leads form near the tidal zone ice. Moving like fianchettoed bishops on a chessboard, hunters hop diagonally across small squares of ice in squelching wellies; their transit to land becoming increasingly more circuitous and implausible by the hour.

In the bay sits a black iceberg liberated in the annual metamorphosis. It must have come into contact with sediment and rock. I lie on the grass with my binoculars wrapped around my neck. Ringed plovers dance the cha-cha by the shore; bar-tailed godwits have upped sticks in New Zealand to come here and watch the show; eider ducks fly over the splintering sea ice; fledgling snow buntings chat to one another about the intruder; flashes of black and white as black guillemots fly past dropping small bombs of guano.

Closely related cousins are on the menu later in the day. Ibbi invites me to his cabin to eat boiled little auks (*amiliq*). How could I say 'no'? Will I ever again receive such an invitation once I leave this place? As always, Ibbi's face shows welcome. He is soon busy knitting the entire world into some kind of conspiracy theory. The kitchen is cluttered with dead eider ducks, gulls, cormorants and guillemots. A harp seal on the floor. A taxidermist's dream. Like an ornithologist anorak, I have been lying camouflaged in the long grass all day jotting down bird names when I could have just come here and done the identifying:

'Bang, bang', shouts Ibbi pointing his rifle up in the air imitating the shooting of ducks. I try to share his enthusiasm, but am much happier watching it all through the lens of my binoculars. We spend the whole evening talking about the changes in the ice, the liberation of the waters and what it will all bring in terms of birds, sea mammals and fish:

'Stiffi, we live from all these birds and animals. They come here and it is our food. But now there are far fewer birds and ducks than before. Everything is changing', says Ibbi. It is a familiar story. And it is not just about numbers. The need to live in harmony with the external world is well-understood here. Can we achieve health or happiness without this? OK, enough of the clichés.

The change in the seasons really means something though. Proof that the planet is still working. It's like some kind of resurrection. There is renewal and newness everywhere. I feel it in my body and mind. I don't want to be out of kilter with all this creation. My diet changes; my sleep patterns change; even if I lack their ancient instincts, I try somehow to live alongside the animals and the birds. This non-human world has leaked into mine.

I return to the cabin, happy in the thought that I am liberated an incy-wincy bit from the man-managed world.

June 7

I day-dream through the morning. The light is causing havoc. The sun rotates above me, assaulting me at any time of the 'night'. The temperature ticks up day-after-day. Shadows flying from their arms and legs, children now run around in T-shirts and I have discarded my coat. My cabin has five windows, and by mid-afternoon is turning into a little furnace.

One-by-one I open the windows and admit the bird song. I let the world in. I scan the tundra and the ice. The sea ice is becoming quite thin in places, and so halibut fishing is coming to an end. The occasional woman with a spherical face still fishes through the leads. Under a flawless, empty sky, I go to greet them, and then hopping across the pancake ice I follow in the tracks of the snow geese that have blown in with the south-westerly winds. They settle for a while, make a tremendous din on arrival, and then get down to business (they are not here for the sake of it)—the grubbing for the roots of sedges and grasses in the boggy area—before moving slightly further along the shore to a spot where the ice is buckled. I can relate to their restlessness. I follow them as far as I can, but then give up the chase when they move inland and out of sight. Their honking slowly fades.

With the calls of geese ringing in my ears, I head back to the hut across all this implacable grandeur and cook supper. More halibut. After dinner, I write at the table and then switch on the radio: a church service followed by Mozart. It must be another Danish Bank Holiday.

Evening: I lie on my bed reading *A Subaltern's Love Song* by Betjeman. Here, there is little chance of having to worry about the "double-end evening tie" and zero chance of the "scent of conifers", but sometimes I understand his nostalgia and dislike of the modern.

This nostalgia seems to grow slowly in us as we become older. I peep through the curtain to see what has become of the snow geese. They too remind me of an earlier time. I spot their black wing tips flicker over the tundra. As confused as I am, they do not roost but appear set on feeding all night long. On the other side of the hut, the phalaropes and sandpipers are out gallivanting, looking for a mate or two.

The constant light is disorientating. I know that must be beginning to sound like a cliché. I try to sleep, but it is impossible even with my two blindfolds. Here, at the top of the earth, we are tilting towards the sun and there is little I can do about it. The Inuit have ditched their circadian rhythms, and so I do the same. It is two in the morning. I get up and head off to join the lonely figures on the ice at the end of my binoculars. With all the meltwater, the tundra is turning into some kind of bog. *Nipi* and *Amak*—my accomplices—follow me like dutiful footmen. Perhaps out of boredom or sheer curiosity. Given up on sleep, I spend the night fishing in the open leads for arctic char. The dogs curl up next to me. It is so bright I have to wear sunglasses. Day, night, whatever you wish to call it, these weeks amount to some kind of delirium.

June 9

If you live alone for long enough in a cabin, your own personality comes to the fore. All those salient features, mannerisms, thoughts and expressions that can only be *you* bubble up to the surface. The rest—those little bits of social conditioning whose life was always ephemeral—wither away.

June 10

It is summer in the Arctic. It is twenty-five degrees in the cabin. Jens Ole saunters past wearing shorts and a T-shirt. This is the madness of the world in a time of climate upheaval. This is all our doing. For how long can all this endure? We are up against the wall now.

* * *

The tundra has been transformed into patches of iridescence. Cotton grass and arctic poppies in bloom. Battalions of little ringed plovers march along the shore. Down on the beach, Rodin's legacy in ice: *Le Penseur, La Danaïde, Le Baiser et La Misère.* Through the binoculars, I can see that the sea ice is broken in places, shattered and scattered like a love story with a bad end.

I sit outside, having my breakfast on the balcony. I scrutinize everything, constantly questioning the world as I find it. The neighboring hunters do likewise. Or, at least I imagine them doing so. We like to take stock of the situation. The tundra is busy with plovers, phalaropes, Lapland buntings and eider ducks. The tension of the light hasn't slackened a bit. The dogs doze under the implacable glare of the summer sun, panting day and night. Having breakfasted, I go and collect meltwater. It is soon pouring off the ice cap straight into my empty buckets. This will be the dog's and my drinking water. It is not long before I have filled up the tank in the cabin. A meltwater lake is forming on the shore ice. I watch all these changes in my natural environment like a hawk. This and the constant sun should melt the remaining sea ice quickly. There are still three or four dog teams on the sea ice. The dogs are further and further out now because the tidal zone ice is all breaking up. To me, it seems like the dogs are going to be stranded on an ice floe, but the hunters must know better. I meet with Jens Ole. He gives me the run-down on the hunting conditions, the mood of his dogs and the latest jokes circulating amongst the bachelors. We study the satellite pictures. There is still in fact about twenty miles of sea ice, but it must be very thin now.

With the shelves at the store more or less empty, everybody is talking about the arrival of the supply ship. She is expected in three weeks' time. At the moment, it will be impossible for her to reach the settlement and berth for she is not an ice-breaker. I am not sure how I feel about welcoming the arrival of consumer goods. I take some satisfaction in being self-sufficient. I love the fact that you can eliminate the intermediaries more or less, and life is no less satisfying. Part of me has warmed to the subsistence life-style to such an extent that I almost perceive the prospective arrival of the supply ship as an imposition, but of course I am just a visitor here and the hunters need their supplies. On the other hand, I cannot wait to eat chocolate again.

June 12

Making sure the hunters are not looking, I have a heart-to-heart with *Nipi* on the balcony. She does not need words. Her affections are inscribed on her face. All that unconditional love. I feel like she has become an extension of me.

June 14

Change is around the corner. Hunters are preparing their kayaks. *Avataqs* which are buoys used on a kayak made from the bladder of a seal can be seen hanging from cabins. Soon, the narwhal will arrive and the hunters must be ready. After polar bear, narwhal is the greatest prize for any hunter.

The morning is devoted to poetry writing. The summer months in the Arctic seem to reveal close-up what we have become. In order to understand your own time, you need to be somehow apart from it. That was one of the objectives in coming here in the first place. A change of perspective can be difficult, but so refreshing. And here I am, looking down on all of this sickness of modernity (partially apart, but not quite) and all that crockery that is going to get broken:

On the edge of a bleak future,
Of water-stricken mega cities and
A mixed-up, globalized world,
The megaphone of universal hip-hop
Blaring to the masses in corrupt English,
Give me The Rolling Stones, Shostakovich and Rach.,
Give me words that stimulate a cabaret of expressions
When thoughts of previous lovers are lost to the wind,
The poet seeks the purity of the tribe,
The nomads of a disappearing age
Who dog sledded in the blurred tedium of the polar desert,
But the vision has long gone,
The last drop of romance has dissolved in the evening sun

June 16

It is that time of transition again. The ice becomes water; my playground melts away. Silence in the cabin. 'A penny for your thoughts'. I am always in transit, always between places and things, normally surrounded by 'strangers'. I want to see everything in the world all at once. My ambition as a child was simple: go everywhere and see everything. But the hunters in this little settlement are wedded to this parcel of hostile land. They will stay here for as long as they are able. And then one day they will be interred in the land. Face up, watching over their kith and kin reenacting the same traditions.

June 17

The hunters say strong winds from the North will blow the sea ice away. But not quite yet, it seems. Alutsiaq is headed to the ice edge with his dog team. His kayak is fastened to the side of the sled. He thinks the narwhal will arrive any day.

I read Swinburne in silence, and trespass into the tranquil dreams of solitude. He suffered from sleepless nights too, but unlike me he was furnished with visions of the goddess Aphrodite:

All the night sleep came not upon my eyelids,
Shed not dew, nor shook nor unclosed a feather,

Yet with lips shut close and with eyes of iron
Stood and beheld me.
Then to me so lying awake a vision
Came without sleep over the seas and touched me,
Softly touched mine eyelids and lips; and I too,
Full of the vision,
Saw the white implacable Aphrodite,
Saw the air unbound and the feet unsandalled
Shine as fire of sunset on western waters

(from *Sapphics*)

In the afternoon, I release myself from one of these moments of withdrawal and visit Pauline up on the hill. I need today to bring something new. Underfoot, the tundra is now boggy. Pauline is a fine woman, a benevolent matriarch, the granddaughter of the polar explorer, Robert Peary, but more importantly a kind and intelligent person: the local guardian of knowledge, traditions and much more. She is a kindred spirit. I knew that from my first meeting with her. She always speaks of how important it is that we experience other cultures. In her book, all outsiders are welcome here and she thinks we can all learn so much from each other. Nobody is excluded. She came on the helicopter from the town where she normally lives. *Akgerit, akgerit* . . . she waves me into the hut where she is staying. Her face grins. 'You must try some Greenlandic food', she says.

We sit together by the sink with a plastic bag on the floor, and she brings a dozen boiled little auks (*amiliq*). She has boiled them in salt for just over an hour and they are now ready to eat. The wings are first removed. The feathers are taken off and the bones are sucked dry. Then, there is the meat on the breast which is a dark brown chocolate color, not dissimilar to seal. The taste is quite similar too, a sort of slightly fishy chicken. It is delicious. The whole bird is eaten: the two livers, the heart, the entrails and the head which is Pauline's favorite is eaten last. She shows me how to suck the juices out of the bird's anus and we lick our lips. We crack the skull of the birds open with our teeth and eat the brains, the tongue and the eyes. We share a wealth of expressions as we work our way through the feast.

I think I have finally gone native.

June 19

Two people with utterly different backgrounds, ontologies, views, tastes and opinions can unite if they share a sense of humor and a light touch for then they can disarm the more combative conversations. That, at least, is one reason to be positive about humanity.

June 20

Days like this are seldom in the Far North. It is overcast, cool and drizzly. The shock of the English weather makes me feel dazed as if I had just awoken from a powerful dream. For the third time in a year, there is rain in our desert. I go outside to feel the first drops on my face. Swirls of mist soon cloak my shoulders. Mist ushers in nostalgia; blurry days in London smudged in my memory. I breathe in all the moisture, and feel for a moment as if I am someplace else. But then I hear the ice moaning: the ice floes are jostling for position with one another, the high frequencies moving faster than the low ones through the ice.

The mist swallows up all the cabins, one-by-one. And so, I return to the hut—the locus of my imaginative life. I read Hilaire Belloc and ponder a bit:

> Of three in One and One in three
> My narrow mind would doubting be
> Till Beauty, Grace and Kindness met
> And all at once were Juliet
>
> (*A Trinity*)

When the world seems to close in on us, we tend to retreat to comfort and security. I wonder. It is an absurdly simplistic taxonomy, but here at least there are 'readers' and 'doers'. I am a 'reader', and I am surrounded by 'doers'. My skills and interests have almost no value or currency here. But is this a *Zen and the Art of Motorcycle Maintenance* (Robert M. Pirsig's 1974 best-selling philosophical travelogue) like scenario? Yes, my attitude to the world is full of romantic viewpoints but the hunters are not slaves to rational analysis. They are in fact alive to the irrational, spiritual world of the supernatural. The 'reader' may be inclined to look back with nostalgia whilst the 'doer' looks upon technology as a means to a better future, but we can both derive pleasure from *sled maintenance*. I am motivated to repair the sled because I wish to perpetuate the romantic vision of travel without machines, whereas a 'doer' can get pleasure from the mechanical process of fixing the sled in the knowledge that it will run better. My feeling is that 'we readers'—those blurry-eyed romantic spirits who eke out solace in the corners of the earth—are becoming a minority; lonely voices on the periphery of a rationalist, technological and scientistic world. I do not wish to be a lone voice amongst logical positivists. Here, I have found a welcome break from that kind of mindset. A life less dull and more captivating for there is still scope for intuition, wonder and surprise. Our great 'conversions' in life normally come about through intuition, flashes of transcendentality, inexplicable moments of aporia and not actually intellectual debate.

Thought this evening: when we lose somebody very close to us, they can still live on in us through their language that we subconsciously embody. That is the case with my grandfather with whom I was so very close. I have absorbed so many of his personal idioms, expressions and turns-of-phrase. These bits of language belonged to him, and now through my usage of them part of him lives on in me (if that doesn't sound too biblical).

Chapter 5: *On Valediction*

Never to bid good-bye
Or lip me the softest call,
Or utter a wish for a word, while I
Saw morning harden upon the wall,
Unmoved, unknowing
That your great going
Had place that moment, and altered all

(*The Going* by Thomas Hardy)

June 22

Poems came to Robert Browning in his dreams:

This was my dream: I saw a Forest
Old as the earth, no track nor trace
Of unmade man

(*Bad Dreams III*)

Mine came to me in the polar darkness, but are now beginning to fade. Instead, my thoughts are turning inexorably to my departure, the world I will be returning to and the whole plurality of humanity:

Thinning candleflames flicker and whisper,
On evenings that are self-doubting,
Amidst all the *thèmes du jour*
That are not discussed,
The changing *Landsgesicht,*
And the silence of my political coevals,
And the loss of iconicity,

The never-ending rain that
Sings a morning madrigal
Running down the gutters of steep streets,
And the puddled pavements glazed
In the light of nostalgic, bow-fronted shop windows,
Now that my upholstered thoughts turn to home

Thought this evening: if you think something is wrong, you have to speak up. If you have the courage to do so, others will follow suit. You can join the bandwagon and say nothing, but it might end up with the Holocaust. That is what happened in Germany and Austria in the 1930s, and it could happen again. As Blake said in *The Marriage of Heaven and Hell*: "Always be ready to speak your mind, and a base man will avoid you."

June 24

The close of evening. A local woman who has apparently taken a liking to me walks into the hut and asks for tea. It is not long before she starts bothering me to extend my patronage further.

June 26

There is still one and a half feet of sea ice, but here on land we are sweating in the heat. The dogs' collective panting sounds like some kind of throat singing. The panting seems to only cease when they are being harassed by ravens.

In the evening, I am out walking again through places of vast horizons. I walk far along the shore. The ravens gargle and croak as they perform aerial acrobatics above me. The earth appears to tremble as heat rises from the stones. I photograph lichen covered stones with scores of different smiling faces. Between pools of meltwater and sat on a carpet of wild flowers with my binoculars, I chew on moss campion and watch the *uuttoq* ('seal or walrus that has crept up onto the ice to bask in the sun') and the occasional black-throated diver. The tundra is like a cushion. I stare into infinity. This endless horizon seems to magnify my sense of freedom. There are a number of hunters out on the sea ice and as many as seven dog teams parked the other side of the tidal zone ice. They hop from one bit of floating ice to another in their wellies, shouting commands. It is not uninteresting to watch them go about their business.

Further along the shore, I find eider duck eggs. I put a couple into my pockets and retreat to my cabin for supper: soft-boiled, the yokes are creamier than I have ever tasted before. This may seem like a harsh land, but there is everything here. We don't need all the mindless consumption stuffed into mountains of non-biodegradable plastic containers from the 'other world'. We don't need it at all.

Note: the ethnographer, Knud Rasmussen, was here almost exactly one hundred years ago and said in the preface of his book to the Polar Eskimos that if anybody wants to study these people, they need to come as soon as possible as they are so few, they might not be around for so much longer. How wrong he was. Today, there are more Inuit living in this part of the Arctic than ever before. For decades, climate scientists have been saying that their way of life, travelling by dog sled over the sea ice, will have disappeared within ten years. It is almost July and the Inuit are still travelling with their dog teams on the ice. Dramatic changes are taking place, but it is not all going to disappear just like that.

June 28

The sandy tracks are alive with snow bunting chatter. They nest under the small rocks. Even though they are immediately beside the cabins, most of us are oblivious to their presence. I am minded of the fact that there is so much going on around us that we are quite unaware of. In front of the settlement, there are now two wide pools of open water. One of the hunters is using a visiting motor-boat to ferry in his dog team. To the west, some families are camping in tents on the tundra, hunting ducks, making tea from saxifrage and living the old peripatetic life. I join them for a while. The men are barbecuing caribou and everybody is in high-spirits. Then, I go looking for crowberries to keep the scurvy at bay. This amuses them for berry-picking is considered 'women's work'.

It is not long before I am eating crowberries and chewing on the stems of mountain sorrel. I value these spontaneous experiences. This is an education too. Lunch complete, I lie down in the spongy tundra canopied by sedge, cotton grass, willow shrubs, mosses and lichens and look up at the superb sky. This slow life and the non-stop sunshine make me wonder sometimes if the world has stopped rotating on its axis. Perhaps God has closed up shop, but forgotten to tell us few 'up here'.

Note: It might be the pressure of time that distills beauty, but it is the total lack of time pressure that enables us to appreciate the simple pleasures.

Memo: to me, an understanding of how to live sustainably in your environment is *real* knowledge. Not physics, technology and all that stuff. If we were to value technological advances in terms of their contribution to our spirituality, their value would be close to zero. These 'advances' are just fancy trimmings. The older Inuit hunters know how to live and survive in this place without recourse to western consumer culture. That is what impresses me. What I like about the older generation is that they learnt the laws of human life and toil from the Arctic environment itself. The consumer culture that has been unleashed on them just dulls the mind and alienates you from your local eco-system. We just drown in heaps of plastic containers.

If we treat consumer culture like an absolute truth, we begin to die as intelligent human beings that are curious and ask questions about our world. Our actions, talk and interactions with our environment become robotic and hollow. However, if you are ingrained in the local ecology, you understand the complementarity of everything, the complexity of relations and your imagination is intact. We don't need any more knowledge. We have far more knowledge than any of us know what to do with. What we need is wisdom, and that, it seems to me, is becoming an endangered commodity.

Before I fall asleep, I read *A Death in the Desert* by Robert Browning. In this philosophical poem we are reminded that even though man is apparently becoming ever more knowledgeable, he also risks straying into the path of ignorance for he has lost his faith in God and instead places all his trust in material proofs:

> "I say, the acknowledgment of God in Christ
> Accepted by thy reason, solves for thee
> All questions in the earth and out of it,
> And has so far advanced thee to be wise.
> Wouldst thou unprove this to re-prove the proved?
> In life's mere minute, with power to use that proof,
> Leave knowledge and revert to how it sprung?
> Thou hast it; use it and forthwith, or die!"

June 30

Cicero said that philosophizing is nothing other than getting ready to die. Me, instead, I sit here doing all this philosophizing because I want to know *how* we should live.

July 1

Ane-Sofie sits at the table. She is upset for she has heard that I will soon be leaving. She is already counting down the days.

The cabin is too hot. We go outside. Her head supported on her fist, she sits with me on the balcony, drinking endless cups of tea. She helps me pronounce the local words (for example, *pilugginnangitsorruanga* 'I am not joking'). It sounds like the wind blowing down my chimney on one of those dark, winter nights. For a moment, it is difficult to believe that I spent months in this cabin not knowing if I would make it through the winter or not without losing a digit or two. Now, the sweat drips from my brow. The inner and outer metamorphosis is complete. Summer buzzes around our heads. The air is loaded with insects: blue bottles, lady beetles, bumblebees, a northern clouded yellow butterfly and the first mosquito. Everything has hatched. Meltwater pours off the ice cap behind us. There are now large areas of open water around the remaining icebergs. Nature has come to us.

July 2

A 'northern clouded yellow' flies into the cabin. I feel as if its entrance must symbolize something. I have a Nabokovian fascination for these remarkable creatures, and follow its passage around the hut. Their hybrid nature, their satin wings, their ephemerality. They have just one objective in their short lives: find a mate and allow the cycle of life to continue. They live for just a couple of weeks. Like life itself, we are touched by beauty and then it is gone.

July 4

The last day of hunting on the sea ice. The remaining dog teams are all being brought ashore. The work is shared. The dogs hurry across the last fragments of ice floes occasionally crashing into the cold water.

The fleet is soon in. Under fitful bursts of sunlight, the dogs chew on dried halibut on the tundra. Everybody watches the proceedings. Nature's spectators. The hunters—the kings of this world—pat their full stomachs, content with the results of this dog sledding season.

Before me, almost impossible images that should be held in trust for the common interest. The sea ice is striped dark blue and shades of jade where the meltwater has mixed with the glacial ice and then refrozen; small clouds of *pujoq* sit above the bits of open water that are now fully open and not just surface meltwater. It is only when I look at such vistas that I realize how we have grown so dangerously abstract in relation to what our actual human needs are. Observing nature in this way enables me to go back to a time when aesthetics was still part of our daily life.

A refugee from 'clock time', I realize that if I blink I might miss the summer. Everything is changing at such a tremendous pace. In this ephemeral Arctic world, the odd wild flower is already beginning to die. Now that some of the technology that stands between me and the creative process of life has been removed, I can at least get to grips with natural time. I can see the summer will soon end because the saxifrage is wilting and not because it is the beginning of July. These little connections enrichen the experience of the world. When you get rid of the structures that contradict time, you move a little closer to the essence of life.

July 6

The remains of the sea ice wreckage have almost gone and with it our highway will soon be gone. Once again, we will become two worlds: land and sea.

July 7

Just as the hunters predicted, strong winds from the North blow the remains of the sea ice away. It is slightly strange to see the open water again; the blue replacing the white. It has been almost ten months. My mind had decided somehow that the ice might be permanent. I will now have to travel another way.

I go for a long walk in the evening to see what has become of my view. Within twenty-four hours, almost all of the sea ice has disappeared. As with everything in the Arctic, change is fast and dramatic. The wild flowers are the fastest growing in the world and even the dogs seem to grow up unfeasibly fast.

July 9

The weather has turned moody. I dedicate the day to poetry and vagabond thoughts. My one and only volume of poetry—that delicious tome of word-smithery—lies open across my knees for long periods of time. Centuries of introspection, somehow compiled into one great book weighs down on me:

> He is gone on the mountain,
> He is lost to the forest,
> Like a summer-dried fountain,
> When our need was the sorest
>
> (Sir Walter Scott, *The Lady of the Lake*)

Those traditional ballads try to sing in my mind, but I am brooding. My mind assembling adjectives (natty; beef-witted; crapulous; sassy; blithe). Now that I am coming to the end of my stay, I have started to moralize:

> And soon time to leave my *Elsewhere*,
> Time to return to my *Lebenswelt*
> Of chloroform, neutered thought,
> To the hullabaloo of the circus,
> To a different kind of *Mitsein*,
> To the hubbub, hurry-scurry and helter-skelter
> Of the minutiae of modern life,
> Of the millions and billions,
> To cramped, urban sprawl-sopped England,
> Packed in their tiny, terraced houses,
> Many-hued faces, faiths and fancies
> Upon England's lost mountain's green,
> Sitting in their identical motorcars,
> With idle expressions and bad postures,
> And a roster of dilemmas,
> Waiting for scuttled hopes and scotched futures
> To close clotted narratives
> That prowl in din-dinted parlors

July 11

A poverty of imagination confines me to the cabin. It is eighteen degrees Celsius in the High Arctic. To me, it feels more like thirty degrees. The hunters say it has never been this hot before. Ibbi is visiting and, as usual, is prepared to concede very few superiorities. In between singing short ditties, he reminds me:

> 'The ice will not disappear. This is a warm period which will last for thirty years or so, then it will turn cold again.'

Ibbi repeats his theory that Earth has moved closer to the Sun and that this is the reason for global warming. After each such statement, he laughs and shouts *uanga angakkoq angihoq* ('I am the big shaman'). Let's see if the 'big shaman' is right or not.

July 13

The supply ship, *Arina Arctica*, has arrived. The horn is blown four times; children run down the grassy slopes towards the shore. Parents sit at the window with binoculars in hand. There is no need to rush. It will be a day or two before all the goods are unloaded. My days of subsistence will now come to an end as the store fills up with Danish produce. I like to think I might carry on eating my raw fish and eider duck eggs, but I know the temptation of chocolate and condensed milk will prove too much. I have been unimaginably frugal these last few months, but somehow it has not felt like a burden. On the contrary, the satisfaction of relying on just nature has given me a spiritual like contentment. More is not necessarily better. This life style has perhaps also contributed to the fact that these memories will be unforgettable for they are simple and decisive.

If you have gratitude, if you can become truly grateful for all the small things you have in life, then it becomes easier to curb your desires. Instead of chasing material goods, if you can find satisfaction in what you already have, then you are surely in a better position to weather any tough times that might lie ahead. And if you live simply, then a gift can become something really special.

Note this evening: if we live in a society of consumer goods where everything is available on the metaphorical supermarket shelf, might this blunt our sexual desires? The *conquest* of hunting has been sexualized for such a long period of time. If the element of conquest is removed entirely from our day-to-day needs, are we left with a kind of impotency? The problem with sitting up here in the cabin is that you live in the Heraclitean shadows. No matter how abstract and preposterous your thinking, you are convinced that somehow your musings are doing a service for the grandeurs of human destiny. This is the result of living a simple life in a remote place that *seems* 'apart' from modernity.

July 15

I am alone in the hut with my Swedish chocolate, my memories and my reminiscences made out of unfinished dreams. I revel in the endless play of the dialectics of realities and imagination, the soul and the mind. And in doing so, I am perhaps trying to live in the atmosphere of another time.

Ten things that cheer me up:

- dogs
- the sun
- a view stretching for 100 miles
- the friendly chirp of snow buntings
- fresh snow
- prospect of an adventure
- empty tracks
- Swedish chocolate
- skies without contrails
- the sound of the chapel bell

July 17

My old world is encroaching on me: the Queen of Denmark has arrived aboard HDMS *Ejnar Mikkelsen*—the Danish Navy vessel that patrols Greenland. A dinghy approaches the shore: a group of serious looking men do the obligatory security checks. It should not take long: this is one of the smallest and most remote settlements in Greenland. The men all have cigarettes hanging out of their mouths and do not look like a very serious operation. Subsequently, a second dinghy arrives. It struggles to get through the broken-up ice. It is looking embarrassing for the Danish Navy. The Queen eventually arrives, wearing a wet-suit and layers of thick protective clothing. She hops out of the dinghy and wades through the icy waters as if this were her daily routine. A group of people in traditional dress line up to meet her.

The Queen makes a very short speech telling us how happy she is to be here. The Prime Minister of Greenland who is accompanying her acts as her translator. There is some drum-dancing. A choir of five sing a local song. The Queen is given a variety of presents and then we all sit down and eat cake. Then, it is time for her to leave. A slightly surreal morning.

July 19

It is twelve degrees Celsius, but feels twice as hot. The explanation: we are living in a desert. There is almost no humidity, and this week at least seldom any clouds.

I meet Titken down at the shore. As always, he is taking himself very seriously and I can never take anyone seriously who takes themselves seriously. He is in one of his angry moods again. He is talking again about 'how he will shoot Muslims if they come here', and asks me constantly why Europe 'is committing suicide by letting so many in'. My mind is on other matters. In the far distance I spot what appears to be a skua: whitish belly, grey wings and a short bill. I follow the bird and soon end up on the melon-scented tundra (from the bog bilberries) photographing arctic rhododendron, mountain avens, dwarf willow and river beauties. The grass close to my hut is now two feet tall. I feel as if you could practically sit and watch it grow.

The noises on this summer afternoon: squabbling arctic terns, exploding icebergs, Greenlandic rock music, small outboard engines from visiting boats.

By the shore, women are hanging up narwhal skin to dry. I watch, and wait for the tide to turn. Soon, my tide is going to turn and I am getting mentally prepared for it. I have no choice now as I am constantly asked: 'why are you leaving?'

Memo: it is often said that you should judge people by what they say about others. I am inclined to agree.

July 20

So, it is the summer season. That means it is time to leave the dogs behind and go paddling in kayaks. But here there are great dangers too. The Inuit cannot swim for there is nowhere to learn. If you capsize the kayak (very easily done) and you cannot roll it, then you drown. Ibbi once told me in a matter-of-fact way that even if you could swim, it would only take ten minutes to die in these waters. Then, seeking agreement he raised his eyebrows.

July 22

Ibbi and I are out hunting narwhal. He has borrowed a speed boat from a hunter that is visiting from the town. When I say 'hunting narwhal', I mean we are sitting in a boat in a distant fjord sharing jokes and occasionally looking overboard to see if there is any sign of these mythical creatures. A cigarette dangles from the corner of his mouth. He pours himself one coffee after another from the Thermos. Glaciers grumble and moan in the background. A one hundred feet wall of ice that stretches from one side of the fjord to the other that reminds you we are entering another Age. Everywhere you look glaciers are retreating up the ice sheet.

Occasionally, a seal pops its head above the water. Ibbi shoots at it, but the seal is quicker and nimbler than him. I am quietly relieved. Through the binoculars, I scan the horizon with glorious images of sparring narwhals in my mind. I cannot imagine hunting one of these mammals. I just want to see one. More chatting and laughing. We snack on *mattak* and dried mackerel.

Late afternoon. I have spotted something not far from our boat. I hear a whale come up to breathe to my right, but do not see anything. A whiff of krill. Then, Ibbi catches sight of it. We sit silently, but the narwhal seems to have moved on. Drama complete. I make do with watching a black-throated diver instead. Slender; magnificent; dressed to impress:

'Stiffi, I admire your patience', Ibbi says. 'You have become one of us!'

Some hours later, an unexpected fog puts an end to our laziness, and we dawdle back to Savissivik, zig-zagging through icebergs in the mist. Kittiwakes and fulmars escort us home. The settlement is deserted. It is summer and the hunters have returned to their nomadic roots.

Some new words:

i'ddarggauho 'he laughs easily'

huguggaileqitsanngaaleqihunga 'I am a little bored'

angu'ddaktaaqtoq 'he catches a seal now and then'

Dramatic climate change has resulted in unexpected migration patterns for the sea mammals making them harder for the Inuit hunters to find

July 23

The fog is gone. The sun revolves around above my head. It is midnight, and my face is beginning to burn.

July 25

Four New Zealanders have arrived on the helicopter. They have flown up here today from Kangerlussuaq in southern Greenland. They are working for the North American Minerals Institute and are here to do geological surveys from the air. Their plane is at Qaanaaq and the appear to be here on some kind of recce. They look about eighteen years old and travel the world the whole time doing this geo-magnetic work. They have no idea what they are looking for or who they really work for. Mining companies are employing the most naïve people they can find and for good reason. They do not ask questions.

It is the most random encounter possible. I am deeply concerned that all this mining research is going on. The Arctic should be something resembling a wilderness, but I cannot tell these clueless boys that. They are just kids who are seeing the world and having a good time.

July 27

The 'white diamond' is dazzling in the light. I keep telling myself this is the most magnificent iceberg of all, and it is the view from my window (but I know I must have thought that a dozen times). This gemstone of ice has everything: carat, cut, clarity and color.

Hunters paddle in kayaks. They are almost silent. They glide in the stream and survey the water like herons.

On the tundra, women are arched over picking bog bilberries. This is exactly what they were doing when I first arrived. I have gone full circle and it will be soon time for me to leave. I can now focus on little else.

In the evening, I invite Ibbi to join me for dinner. There is something I want to get off my chest. We feast on seal ribs and boiled little auks and guillemots. Then, after several beers, I touch on what is for me the most sensitive subject of all. What will happen to my dogs when I leave? He raises an invisible rifle to his right eye, shouts 'bang, bang' and then starts laughing raucously:

'No, no. I am joking, Stiffi. Don't worry. I have sorted this out. Naimmanitsoq and I will take the dogs. We have been watching you. You did a good job training them. Normally, we refuse to take dogs owned by *kad'luna*, but you are alright, *kammak*.'

I could almost collapse with relief.

Memo: what is the purpose of all this? Sitting up here, aloof from the world watching the months roll by. To come around to the beginning again and see if I have learnt anything new? To witness first hand all these upheavals that everyone is talking about? Alarming as they are, we all know one day school boys will yawn whilst reading their history. Death and rebirth. Past and future. Life will go on, and my happiness will remain.

July 29

The overcast weather continues, but it brightens up in the afternoon. I have bit of a lie-in as I have nothing planned for the day. The pedestrian pace of life has seeped deep into my veins. Weeks go by and you achieve nothing, but it no longer matters. How will I reintegrate into a world of deadlines and appointments?

I miss a couple of calls from Ane-Sofie, but am unable to call her back because there is no phone network for much of the day. In the settlement, the talk is of who has caught a narwhal. It means a lot for a boy to catch his first narwhal. A kind of puberty rite.

I have been thinking about how I want to say 'goodbye'. I considered hosting my own *kaffimik*, but I am not very keen on *kaffimiks*. I find the atmosphere rather like a wake. They are formal and there is a sense that you come in, have something to eat and drink, say a few words and then are on your way. There is no guarantee either that the people I would want to turn up, would be able to make it. I think I might just go and visit those I have really got to know.

July 30

I spent the evening drinking with Ibbi. He was busy plotting my future. He refused to give an inch, determined in his belief that I will settle here. I know in thirty years' time he is going to be still talking about his plans for independence for north-west Greenland. Some dreams should never die.

By the time I have sobered up, the day is almost over. Plato forbade people under the age of forty to get drunk. Oh, dear!

July 31

I have almost recovered. Recovered enough to watch the caravan of clouds gliding by. I wonder if I will miss all of this. All this space, time and self-sufficiency.

I make Ane-Sofie dinner: roasted caribou. I can see all she wants is Pizza. She is subdued and upset that I am leaving. She asks me repeatedly when I am leaving. I tell her that I am leaving on Wednesday and she asks how many days away that is. She wants to come with me, but I fear she may be disappointed by the other world she will discover. The next couple of days will not be easy.

In the evening, I pack up all my things and Ane-Sofie insists in helping me clean the cabin. She is loyal to the end. We are a good team, and the job is soon done. Only now does it feel like I really am leaving.

August 1

It is time to say my goodbyes. *Nipi* and *Nukka* already know that I am leaving. They read me like a book. I don't need to tell them anything, but we will have our pull-up-a-chair chats. I have to look after my girls. I'll miss them. They have filled the landscape so completely.

I am just about to leave the cabin when Pauline enters. She has brought me a present: a lace table cloth that she has sown. That was unexpected! Then comes Jens Ole with a Savissivik fifty-year anniversary pen. As with Pauline, there is an air of formality to his visit. He holds forth about how difficult it is to get hold of these pens. And, who am I to disagree? Soon, there is a stern-looking line of visitors outside my cabin. All of them bearing gifts: carvings of kayaks, key-rings, bits of antlers, polished pebbles. A few tears are shed. Not mine.

The visitors dispersed; I make my way down to Ibbi's cabin. Mid-afternoon. Much snoring. Various family members are sleeping on the sofas and beds. We eat frozen mackerel and slivers of frozen seal blubber. Not much seriousness here. We struggle to suppress our laughter, and thus not wake up the extended household. The louder our laughter becomes, the more forceful is their snoring in response. Fortunately, Spinoza said there cannot be too much *hilaritas*.

After a few more goodbyes, I return to the cabin. The world has fallen silent. This is it. My time here is almost up. I wanted to go full circle. I have done that, and now that I have begun to feel the terrors of infinite time, I am ready to come down. I know that I will look back at this year as episodic; sharply defined at the beginning and end but without plausible extension.

As with each evening, I curl up inside the sleeping bag and reach for the poetry. John Donne's *A valediction: of weeping*:

Let me pour forth
My tears before thy face, whilst I stay here,
For thy face coins them, and thy stamp they bear,
And by this mintage they are something worth,
    For thus they be
    Pregnant of thee;
Fruits of much grief they are, emblems of more,
When a tear falls, that thou falls which it bore,
So thou and I are nothing then, when on a diverse shore

Under the spotlight of the sun, I say goodnight to the world.

August 2

More valediction. Displays of humanity. My old friend, Taateraaq, grabs me by the arm. Always grinning. Just like the day I met him outside the church:

> *Naangmanniartutin, naangmanniartutin* 'look after yourself, look after yourself'

I go and find Savfak. I tell her that I am leaving today and she begins immediately to cry. She holds my hand and will not let go. She pulls me close to her. Tears pour down her cheeks. I tell her that I have been out to Herbert Island again. That only makes matters worse. She tells me how happy she was out there, but then her husband fell ill and things fell apart. I feel sorry for her. Tragedy has eaten into her life; her stories all belong to a long since abandoned island whose few remaining homes have been fed to the wind. I need to go. She sits by the window; her tears spill onto the pane.

On the way back to the cabin, I meet Ibbi. He is going out hunting, and says we should say our goodbyes now. It is quick and unemotional. A firm handshake; a sparkle in the eye. He is laughing as always:

> *Naangmanniartutin aggurruaq, kammak* 'take good care of yourself, my friend'

Flanked by *Nipi* and *Nukka*, I eat lunch on the balcony. The dogs are on lookout, and take their duties as seriously as ever. Even if the Master is leaving, they still don't want any intruders.

The view is gripping, little different form the first day I arrived. Struck inarticulate by the splendor of all this. The world and its beauties never grow old. Will I ever find the right words to describe all this? A long thin bank of mist stretches right across the bay. The Gods divvy up the horizon. I am soon joined by Ane-Sofie and Qaaqutsiaq. Then, fog descends on the settlement. It is almost as if the Gods are trying to tell me that I should not leave. Will the helicopter come? Shrugged shoulders.

The fog disappears as quickly as it arrived. Qaaqutsiaq can see the helicopter in the distance. I have no idea how. I need another five minutes to confirm that he is not pulling my leg. Ane-Sofie and Qaaqutsiaq help me carry all my things down to the helicopter pad. *Nipi* and *Nukka* in tow. As always.

I hug the dogs. They push their snouts into my chest, copying and reciprocating my actions. They know what this is all about. I give Jens Ole the key and say goodbye to Ane-Sofie and Qaaqutsiaq. She is brave and does not cry. What a girl. I will miss her cheery smile, her tender spirit. As I walk to the helicopter, she shouts 'Stiffi' and gives me one last wave. I board the helicopter. I am the only passenger. A private journey full of private thoughts. Emotions pulled in all directions. It is always the best and the worst that you remember, not the mediocrities. And, Ane-Sofie, Ibbi and the dogs were always the best.

From the air, the settlement looks entirely implausible. So tiny, so far from anywhere. It is exactly what I went looking for. I still cannot believe that this place exists and the words I have chosen to describe it seem painfully inadequate and do not do justice to the images in my mind. Savissivik soon becomes a tiny speck of humanity, and then it disappears completely from view. I am reminded of how such an experience, a thought or even a possibility can transform us entirely.

The flight from Qaanaaq is on time. It is a two-and-a-half-hour flight to Upernavik where a couple more passengers get on and then another two hours to Ilulissat. We arrive there almost exactly in time, at ten o'clock in the evening. The sun will set in Ilulissat for the first time this summer. It really is time to go home.

* * *

Outside the tiny airport at Ilulissat, passengers are greeted with flies and mosquitoes.

Epilogue

I returned first to Denmark and then England that August. I had spent much time thinking about how I would readjust to my old life and whether there would be any difficulties in doing so. The plane touched down at the Copenhagen airport in Kastrup at nine o'clock on a Thursday morning. It was just getting dark, and after four months of twenty-four-hour sunshine, the return to a more regular balance of day and night was very welcome. With seventy-five kilograms (one hundred and sixty-five pounds) of luggage and a pair of skis slung over my shoulder, I staggered out of the airport and was struck by the stifling humidity. After having lived in a polar desert for a year, the sweat poured off me as my taxi driver negotiated the busy streets of the Danish capital. I dropped my bags off at the hotel, walked down one of the main boulevards and sat quietly on a bench. One step at a time, Stiffi.

Even though I had returned to the world I knew, I felt in a sense displaced and overwhelmed. The sky rumbled with the incessant noise of airplanes overhead, one after the other; streams of traffic poured past me in both directions. So many cars. Where was this flood of steel all going? The pavements were almost empty. The only signs of life were fixed heads behind steering wheels, moving horizontally and swiftly as if on a conveyor belt in the wrong gear on an assembly-line of environmental destruction.

It was the dream of Ane-Sofie Imiina who visited me more or less every day and from whom I learnt much of their language, to leave north-west Greenland and come to Denmark or England. But having lived in the High Arctic, I felt unsure about showing her my world of 'progress', a world of industry, high-rise buildings, noise, polluted skies and rivers, a place where childhoods are spent in the back of cars on clogged streets and where freedom comes at a price. What I learnt from life in the Arctic is that it does not ke this. We can live in sustainable ways which are simple and very ut making the change can seem exceedingly difficult.

Cambridge, I spent the first few days diving in and out of the Sainsbury's freezer aisle on Sidney Street. What I perceived to be humidity seemed to be initially at least unbearable. Occasionally, I would look around for the inflamed eye of the sun but there was just grey cloud. In those first weeks and months, I found trips to London challenging. I remember Ibbi telling me about the first time he went to Copenhagen. He asked his host just to leave him in the park for he could not come to terms immediately with the noise. That was how I felt.

Friends and colleagues were anxious to hear my stories, but they were struck how little I had to say. Whilst they meant well, their questions seemed almost intrusive. I had known a world of silence, raised eyebrows and screwed up noses and now there was noise and intellectual interrogation. Or, so it seemed at first. I would struggle to find the right words, and would sometimes instead just raise an eyebrow. Shortly after my return, I was convinced that nobody saw the world the way I saw it, and perhaps I was right. I had been afforded glimpses of a life that transcended the limits of our perception, and was anxious not to lose those insights.

If you live in the Arctic for a period of time, you can never quite leave it and it never quite leaves you. Two years later, I returned briefly to the community and found that not everything stays the same. Ane-Sofie with whom I was in regular contact moved to Norway to continue her schooling. She soon felt home-sick and returned to north-west Greenland. Naimmanitsoq dropped dead from a brain seizure shortly after my last hunt with him. Ibbi though was still choking on his own laughter. *Nipi* and *Nukka*, thank God, are doing just fine. And, me, well, I thankfully managed to resynchronize. But perhaps not entirely. Even now when I hear a helicopter overhead, I look up longingly and my mind returns to that magical place.

Afterword

The reader could be forgiven for thinking that life in a remote corner of Greenland is some kind of paradise for he or she who wishes to escape the minutiae of modern living. Whilst my year there was enriching in so many ways, I fully appreciate that life is not always easy for the last remaining hunters living in the tiniest of settlements. Alcoholism has plagued many of the communities, suicide rates are amongst the highest in the world, more recently many have fallen ill to cancer—a disease that previously had very little impact on hunters. A sudden change in diet is surely the explanation for the increase in so many health problems. What is more, a number of hunters still die of accidents resulting in the fact that few live beyond the age of sixty-five. I am fully aware of all these things and all the related social issues, but choose not to discuss them here for this is meant to be a book about the things I love.

Glossary

ajor	oh no, oh dear
ajorpoq	'bad' or 'not functioning' (strictly speaking, the word is West Greenlandic but is sometimes used as a synonym for *naamaangitsoq*)
ammaqa	perhaps
angatsuduk	bachelor
hamani	'down there', used in reference to the 'other world'
hiku	the sea ice
imaq	open water
inuk	singular of Inuit
kad'luna	the white man, often used to refer to the European
kaffimik	a party held on special occasions to mark birthdays and other events. Sweet and savoury food is always served and coffee is drunk
kammak	friend
kamikker	seal skin boots lined with arctic fox or arctic hare fur and worn by the Inuit
kiffaq	previously, a woman that would do chores around the house
kivittok	a feared, semi-mythical figure that goes out up into the mountains or onto the Ice Sheet because he has been rejected from society

mattak	the skin and blubber of a narwhal. A delicacy in Greenland, high in Vitamin C
nalorrhorruiga	'I don't know'
nannut	polar bear fur trousers worn by hunters
nigeq	a strong wind that blows from the East in the months of March and April
nuna	the local cosmos, the land, sea, sea ice, air and all the stories, feelings and beliefs connected to the natural environment
pujoq	smoke or cloud, in particular puffs of cloud that appear over open water
qujan	'thank you'
taima	'enough', often used to end a conversation
tarhaq	a white shield that hunters hide behind when they shoot seals basking in the sunshine on the sea ice
tarneq	the soul of a human being or animal
toorngaq	the spirits that would follow and help a shaman in time of need
uniit	'never mind'; 'don't worry'

Lightning Source UK Ltd.
Milton Keynes UK
UKHW021827111022
410321UK00008B/138

9 781666 738285